Wine Traceability

Wine Traceability

Special Issue Editor

Maria Carla Cravero

MDPI • Basel • Beijing • Wuhan • Barcelona • Belgrade

MDPI

Special Issue Editor
Maria Carla Cravero
CREA Council for Agricultural
Research and Economics
Italy

Editorial Office
MDPI
St. Alban-Anlage 66
4052 Basel, Switzerland

This is a reprint of articles from the Special Issue published online in the open access journal *Beverages* (ISSN 2306-5710) from 2018 to 2019 (available at: https://www.mdpi.com/journal/beverages/ special_issues/wine_traceability).

For citation purposes, cite each article independently as indicated on the article page online and as indicated below:

LastName, A.A.; LastName, B.B.; LastName, C.C. Article Title. *Journal Name* **Year**, *Article Number*, Page Range.

ISBN 978-3-03921-768-7 (Pbk)
ISBN 978-3-03921-769-4 (PDF)

Contents

About the Special Issue Editor

Maria Carla Cravero is currently a researcher at CREA (Council for Agricultural Research and Economics), Research Centre for Viticulture and Enology (Italy). She graduated in Agriculture and post-graduated in Viticulture and Enology (University of Turin, Italy). Her research experience is in grape and wine physical-chemical analysis (1988–1997), and from 1997 to the present day she has studied the sensory analysis of grape, wine, and alcoholic beverages. From 2005 to 2015 she was a member of the Italian delegation to the International Organization of Vine and Wine (OIV), and she contributed to the review document on the sensory analysis of wine. She is a reviewer of some international journals (https://publons.com) and co-author of one patent and the author/co-author of more than 200 papers (19 International peer-reviewed papers, 49 papers in conference proceedings, 120 papers in professional journals and magazines, and 19 monographs or chapters in monographs).

She contributed to translate the "Traité d'oenologie" of P. Ribéreau-Gayon, D. Dubourdieu, B. Donèche, A. Lonvaud, from the French to Italian: II edition (2003), III edition (2007), and IV edition (2018).

beverages

MDPI

Editorial

Wine Traceability

Maria Carla Cravero

CREA Council for Agricultural Research and Economics, Research Centre for Viticulture and Enology, Via Pietro Micca 35, 14100 Asti, Italy; mariacarla.cravero@crea.gov.it

Received: 28 August 2019; Accepted: 29 September 2019; Published: 9 October 2019

This Special Issue concerns wine traceability, a central theme in the current world market where consumers are increasingly demanding the quality and origin of food and drink.

The wine production chain and wine composition are generally controlled by different laws (International Organization of Vine and Wine (OIV), European union (EU) and national governments) and need specific documentation. Nevertheless, wine production is subject to fraud. As a consequence, the improvement of the methods applied to verify the origin and the quality of wines is really important to protect wine consumers and producers.

In this book eight different papers—six research papers and two reviews—address the topic from different points of view.

The first review [1] regards the latest trends in the polyphenol fingerprinting of red wines in association with their evolution during winemaking and storage. Polyphenols profiles are important to characterize grapes and wines, in particular color and sensory parameters, but also nutritional quality. The authors cite the literature on the characterization of the polyphenolic compounds from grapes and wines, polyphenol fingerprinting and analysis in red wines. Moreover, they also consider the current research on the factors affecting polyphenol content and distribution in wines.

The other review [2] presents several strategies to ensure the authenticity of wines, identifying new technological methods that can be employed in wine traceability. The authors emphasize an integrated approach to verify the origin of wine. In the paper we found an overview of the literature on analytical techniques for tracing the geographic provenance of wines. Moreover, the authors focus their interest on DNA fingerprinting for varietal identification and they show the pros and cons of DNA-based techniques applied to wine authentication.

The research papers propose different chemical instrumental analysis together with multi-variate statistical tools as methodologies to verify the origin of a wine. A group of Brazilian researchers [3] proposed a method based on data mining and predictive analysis to classify Brazilian and Uruguayan Tannat wines. In particular, they used the support vector machine (SVM) classification algorithm with the radial basis kernel function and the F-score feature selection method. The variables analyzed in the wines were color CIELAB parameters), total polyphenols, total anthocyanins, antioxidant activity by oxygen radical absorbance capacity (ORAC) and free radical scavenging activity (DPPH), and individual anthocyanins. The authors conclude that given the use of at least one anthocyanin (peonidin-3-glucoside) and DPPH, the Tannat wines can be classified with 94.64% accuracy and Matthew's correlation coefficient (MCC) of 0.90.

A research paper from Portugal [4] studied elemental composition with the aim of discriminating wines according to geographical origin, considering the effects of soil, winemaking process and vintage. The elemental composition of soils, grapes, musts and wines was determined by using inductively coupled plasma mass spectrometry (ICP-MS) followed by multivariate statistical analysis. The study evidenced that the mineral composition pattern is transferred through the soil-wine system, and differences observed for soils are reflected in grape musts and wines, but not for all elements. Li, Mn, Sr and rare-earth elements discriminated wines according to vineyard. Results suggest that winemaking processes and vintage year should be considered if we want to use the elemental composition for wine traceability.

Another study [5] proposes abstract liquid chromatography coupled to high-resolution mass spectrometry (LC-Q/TOF) to identify grape secondary metabolites that can potentially be used as chemical markers for the traceability of corresponding wines. In this experience the profiles of flavonols, flavanols and flavanones, glycoside terpenols, procyanidins, stilbenes, and anthocyanins of four Italian varieties (Corvina, Raboso Piave, Primitivo, and Negro Amaro) were determined, and the peculiar metabolites—the possible traceability markers of the corresponding wines—were identified. The authors suggest that this approach can successfully be used for the traceability of other wines.

Other chemical markers can be considered for wine traceability, in particular, isotope ratios of bio-elements (13C/12C, D/H, 18O/16O) or radiogenic heavy elements like 87Sr/86Sr, which were studied in the Chianti Classico production area [6]. This study aimed to verify whether the Sr isotopic ratio coming from vine branches can be a more distinctive geographical traceability tool. The strontium isotope ratio data (87Sr/86Sr) were acquired using a double-focusing multi-collector inductively coupled plasma mass spectrometer (MC-ICP/MS). The results showed that the 87Sr/86Sr ratio is not influenced by soil particle size and may represent an effective tool as a geographic provenance indicator for the investigated product.

SNIF-NMR (Site Specific Natural Isotope Fractionation, Nuclear Magnetic Resonance) and IRMS (Isotopic Ratio Mass Spectroscopy) techniques are widely used to assess the geographical origin of food, beverages and wines. These instrumental analyses were applied by other researchers [7] in some Italian wines (Verdicchio, Fiano, Refosco dal Peduncolo Rosso and Nero d'Avola). The wine isotope results (13C/12C, D/H, 18O/16O ratios) were subjected to multivariate statistical analysis, such as principal component analysis (PCA), together with wine chemical and sensory parameters to characterize the vintage and identify the origin of some Italian wines produced from different varieties. Moreover, the results of the PCA on all the wine parameters (isotopes, chemical and sensory data) showed that it was also possible to discriminate wines of the same variety produced in regions with different soil and climatic conditions.

The last paper [8] focused on the role of rare earth elements (REEs) as geochemical markers in wine traceability analyzing soil, grapes, musts, and wine samples of an Italian red wine (Primitivo di Manduria) using ICP-MS. The authors considered REE distributions in samples taken at each step of the wine production process and showed that the original distribution in soil remains unaltered in every intermediate product up to and including the grape must. Variation of REE composition can be caused by additives used to promote fermentation or by the interaction with the surfaces of storage tanks.

Acknowledgments: All the authors who kindly contributed with their experience and knowledge to this Special Issue are gratefully acknowledged. All my thanks to MDPI for the opportunity to be guest editor of this Special Number.

Conflicts of Interest: The authors declare no conflicts of interest.

References

1. Palade, L.; Popa, M. Polyphenol Fingerprinting Approaches in Wine Traceability and Authenticity: Assessment and Implications of Red Wines. *Beverages* **2018**, *4*, 75. [CrossRef]
2. Pereira, L.; Gomes, S.; Barrias, S.; Preto Gomes, E.; Margarida Baleiras-Couto, M.; Fernandes, J.R.; Martins-Lopes, P. From the Field to the Bottle—An Integrated Strategy for Wine Authenticity. *Beverages* **2018**, *4*, 71. [CrossRef]
3. Costa, N.L.; García Llobodanin, L.A.; Alves Castro, I.; Barbosa, R. Geographical classification of Tannat wines based on support vector machines and feature selection. *Beverages* **2018**, *4*, 97. [CrossRef]
4. Catarino, S.; Madeira, M.; Monteiro, F.; Caldeira, I.; Bruno de Sousa, R.; Curvelo-Garcia, A. Mineral Composition through Soil-Wine System of Portuguese Vineyards and Its Potential for Wine Traceability. *Beverages* **2018**, *4*, 85. [CrossRef]

Beverages **2019**, *5*, 59

5. Mayr, C.M.; De Rosso, M.; Dalla Vedova, A.; Flamini, R. High-Resolution mass spectrometry identification of secondary metabolites in four red grape varieties potentially useful as traceability markers of wines. *Beverages* **2018**, *4*, 74. [CrossRef]
6. Sighinolfi, S.; Durante, C.; Lisa, L.; Tassi, L.; Marchetti, A. Influence of Chemical and Physical Variables on 87Sr/86Sr Isotope Ratios Determination for Geographical Traceability Studies in the Oenological Food Chain. *Beverages* **2018**, *4*, 55. [CrossRef]
7. Bonello, F.; Cravero, M.C.; Dell'Oro, V.; Tsolakis, C.; Ciambotti, A. Wine Traceability Using Chemical Analysis, Isotopic Parameters, and Sensory Profiles. *Beverages* **2018**, *4*, 54. [CrossRef]
8. Aceto, M.; Bonello, F.; Musso, D.; Tsolakis, C.; Cassino, C.; Osella, D. Wine traceability with rare earth elements. *Beverages* **2018**, *4*, 23. [CrossRef]

beverages

MDPI

Review

Polyphenol Fingerprinting Approaches in Wine Traceability and Authenticity: Assessment and Implications of Red Wines

Laurentiu Mihai Palade [1,2,*] and Mona Elena Popa [2]

[1]   National Institute for Research and Development for Animal Biology and Nutrition,
     Calea Bucuresti Street no. 1, Balotesti, Ilfov 077015, Romania
[2]   Faculty of Biotechnology, University of Agronomic Science and Veterinary Medicine Bucharest,
     59 Marasti Blvd, Bucharest 011464, Romania; pandry2002@yahoo.com
*   Correspondence: palade_laurentiu_mihai@yahoo.com

Received: 31 July 2018; Accepted: 19 September 2018; Published: 7 October 2018

Abstract: Like any other food/feed matrix, regardless of the employed analytical method, wine requires authentication strategies; a suitable qualitative and quantitative analysis represents the fingerprint which defines its identity. Until recently, fingerprinting approaches using liquid chromatography applications have been regarded as an effective tool for the assessment of wines employing polyphenol profiles. These profiles are of considerable importance for grapes and wines as they influence greatly the color, sensory, and nutritional quality of the final product. The authenticity and typicity characters are fundamental characteristics, which may be evaluated by the use of polyphenol fingerprinting techniques. Under these conditions, the evolution of polyphenols during the red wine elaboration and maturation processes shows a high importance at the level of the obtained fingerprints. Moreover, the environment factors (vintage, the area of origin, and variety) and the technological conditions significantly influence wine authenticity through the use of polyphenol profiles. Taking into account the complexity of the matter at hand, this review outlines the latest trends in the polyphenol fingerprinting of red wines in association with the transformations that occur during winemaking and storage.

Keywords: red wine; authenticity; polyphenols; markers; fingerprinting

---

1. Introduction

Nowadays, food quality has become particularly important from the point of view of both the consumers and the manufacturers. Global food safety authorities have generated guidelines that are meant to regulate and assure the authentication and detection of adulteration or incorrect labelling by means of reliable analytical tools [1].

In food manufacturing, traceability is defined as the ability to track a product or product batch with respect to the production history [2]. The associated chain of events covers all of the stages from harvest, transport, storage, processing, distribution, and sales. In various cases, traceability may target a step or series of steps within the production chain [3]. Nonetheless, irrespective of the type of product, traceability is achieved by implementing control systems, which become more complex as the manufacturing steps and the amount of data increase [4].

Traceability in the wine industry, as with any other food product, is enabled by a quality assurance (QA) management system [2]. This implies the need to ensure the availability of specific well documented and characterized documents throughout the process, from grape production to processing and further on to wine distribution [5,6].

When we discuss wine traceability, it is important to consider the socio-economic environment in order to understand how a given vineyard makes progress [7]. Given the fact that viticultural and oenological practices are influenced by mankind, wine is among the food products most subjected to falsification [8]. From the point of view of consumer acceptance, wine authenticity is regarded as having a highly significant implication on wine traceability [5,8]. In other words, the main goals are to combat fraudulent practices, to control product adulteration, and to ensure the organoleptic and nutritional characteristics are valuable for consumers.

As a result of the large chemical biodiversity expressed in plants, phenolics represent an important phytochemical group that is studied with great interest in applied fields such as food science [9]. One of the main challenges in the modern society is the protection of biodiversity in view of a sustainable development [10].

The control and proof of authenticity are generally done by the use of various analytical [11] and statistical methods [12,13], which are based on differentiating the geographical origin, grape variety, wine age, and technology of production [5]. Taking into consideration the scope of the investigation, several instrumental techniques have been employed, such as gas chromatography coupled with mass spectrometry (GC/MS), liquid chromatography coupled with mass spectrometry (LC/MS), ultraviolet-visible spectroscopy (UV-Vis), nuclear magnetic resonance (NMR), near-infrared spectroscopy (NIR), and inductively coupled plasma mass spectrometry (ICP/MS). These methods are widely applied in the study of grape and wine chemistry to certify their authenticity. Moreover, a corroboration of these with statistical methods confers a broader range of confidence within which the certainty of the results falls [14]. In this context, chemometrics techniques may improve the interpretation of the results by extracting useful information and reducing data complexity [9].

The aim of this review was to discuss the advancements and implications of using polyphenol profiles (multivariate approaches) in the endeavor of wine traceability and authenticity.

2. Characterization of the Polyphenolic Compounds from Grapes and Wines

Polyphenols, generally called "phenolics", represent a large group of secondary metabolites with very complex structure, which are essential for the quality and stability of red wines [15]. They have significant importance also in the case of white wines, however, they are present in lesser amounts [8]. Recently, the chemical nature and behavior of polyphenols have gained increasing interest, especially due to the dynamics, concentration, and individual evolution, contributing to the fingerprint of authenticity and typicity of the variety and area of origin [16].

Despite their high diversity and difficulty in being characterized, polyphenols have a common element, which is the presence of at least one hydroxyl group on an aromatic ring (Figure 1) [17]. It can contain a free hydroxyl or another functional group (ether, ester, or heteroside). Given the fact that other metabolites like alkaloids and numerous terpenoids show a benzene ring or a hydroxyl phenolic group in their structure, a simplified definition is insufficient [18]. Therefore, it is necessary to include the biochemical hypothesis according to which, in nature, only plants and microorganisms can synthesize aromatic rings, the building blocks of polyphenols [19].

The following categories of phenolic compounds have been reported in high concentrations in red grapes, musts, and wines: Tannins, anthocyanins, flavonols, dehydroflavonols, and stilbenes.

Non-flavonoid polyphenols   Flavonoid polyphenols

Phenolic acids   Stilbenes

Example: Gallic acid   Example: Resveratrol   Flavonoid skeleton

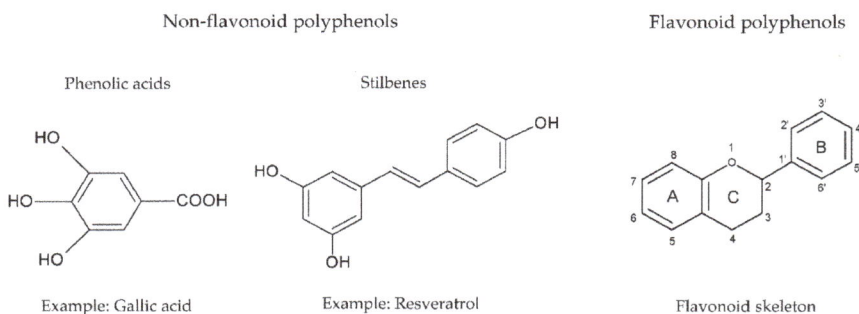

Figure 1. Flavonoid and non-flavonoid polyphenolic compounds. Adapted from References [20,21].

2.1. Classification of Polyphenolic Compounds

In the literature, there are several criteria for classifying polyphenols, however, the most important ones are based on the chemical structure (the association of some compounds that show structural similarity but take into account the high level of complexity) and on the botanical origin (highlighting that some plants produce important amounts of polyphenols, among which the vine is also present) [19,22].

The biosynthesis of polyphenols has been thoroughly discussed [18,23,24]. Generally, polyphenolic compounds are synthesized through two main metabolic pathways, which are the shikimic acid pathway and the phenylpropanoid pathway [19].

Briefly, shikimic acid is the result of condensation and cyclization of phosphoenolpyruvate with erythrose-4-phosphate and is further transformed into chorismic acid. The shikimate pathway, the link between carbohydrate metabolism and the biosynthesis of aromatic amino acids [17], generates phenylalanine which is further supplied to the phenylpropanoid pathway [24]. The condensation of phenylalanine to *trans* cinnamic acid, via phenylalanine ammonia lyase (PAL), is then followed by various interactions with coenzyme A (CoA) to afford the core flavonoid intermediates (chalcone synthase) as well as stilbenes (stilbene synthase) [17,24,25].

2.1.1. Non-Flavonoid Polyphenols

These include phenolic acids, which are separated into two main groups, benzoic acids (C6-C1) and cinnamic acids (C6-C3) (Figure 2), accompanied by stilbenes (C6-C2-C6).

Phenolic acids are present especially in skins but also in the cell vacuoles of grapes pulp. They are colorless in hydroalcoholic solutions but can develop a yellow color when oxidized. They do not influence the sensory attributes of the resulting red wines and show technological importance by being precursors for some volatile phenols produced by microorganisms [18,25].

Benzoic acids differ through the benzene nucleus substituted moieties and are present in grapes in the glycosidic or ester form. The most representative is gallic acid, found in concentrations averaging 95 mg/L in red wines. In turn, syringic and vanillic acids (*p*-hydroxybenzoic acids) are reported in fewer amounts approximating an average of 5 mg/L [25].

Cinnamic acids are generally present in the ester form of tartaric acid and less in the glycosidic one [25]. In red grapes, their concentrations are always higher in the skins than in the pulp. In addition, the content found in red wines is around 10 mg/L, net superior to that found in white wines [20]. It has been reported that tartaric esters of cinnamic acids show a superior capacity of oxidation under the action of tyrosinase, which is naturally found in grapes, as well as laccase produced by *Botrytis cinerea* mould [20]. Ferulic and *p*-coumaric acids can be transformed into 4-vinyl guaiacol and 4-vinyl phenol by cinnamate decarboxylase produced by the viable cells of some *Saccharomyces cerevisiae* yeast strains [20,22].

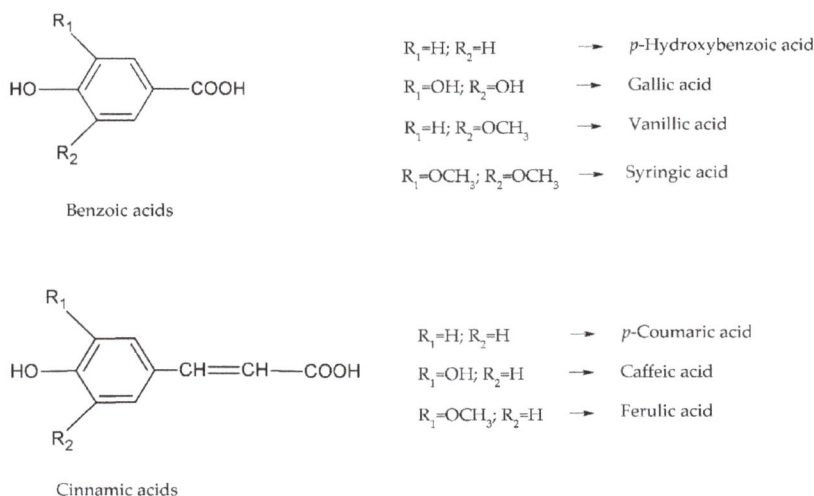

R₁=H; R₂=H → p-Hydroxybenzoic acid

R₁=OH; R₂=OH → Gallic acid

R₁=H; R₂=OCH₃ → Vanillic acid

R₁=OCH₃; R₂=OCH₃ → Syringic acid

Benzoic acids

R₁=H; R₂=H → p-Coumaric acid

R₁=OH; R₂=H → Caffeic acid

R₁=OCH₃; R₂=H → Ferulic acid

Cinnamic acids

Figure 2. Phenolic acids. Adapted from References [18,20].

Stilbenes are a particular class of polyphenols that contain two benzene rings linked through an ethanol or ethylene molecule (Figure 1). Among these compounds, the most representative is resveratrol (3,5,4'-Trihydroxystilbene). The trans isomer of resveratrol is produced in vine in response to fungal infections [17]. It is localized in the skins of red grapes and the extraction of this compound occurs mainly in the maceration-fermentation stage; in the resulting red wines, it is found in concentrations that vary widely from several μg to several mg/L [26]. They do not influence the sensorial and chromatic characteristics of red wines but are important for human health by having the capacity to progressively dissolve the excess fat around the blood vessels [20,27,28].

Phytoalexins, belonging to the group of stilbenes, have recently attracted considerable interest. They are predominantly synthesized in the vine in response to localized stress, both biotic (infections), as well as abiotic, representing a veritable defense mechanism [29]. Piceid, the glycosidic form of resveratrol, as well as pterostilbenes and viniferins, is accumulated in lower amounts [26,29].

2.1.2. Flavonoid Polyphenols

Due to their high variety of structures, flavonoids are of great interest in winemaking [24]. They contain a backbone of 15 carbon atoms comprising two benzene rings (A and B) joined by a heterocycle (C) (Figure 3). Based on the heterocycle, flavonoids can be further separated into three subgroups: Flavonols, dihydroflavonols, anthocyanidins, and flavanols [20,30].

Flavonols and dihydroflavonols or flavononols are secondary metabolites present in almost all higher plants [17]. They contain a pyrone heterocycle and are the second most abundant in grapes [20]. Generally, flavonols are present in both the skins of red and white varieties as 3-O-glycosides (glucosides, galactosides, rhamnosides, rutinosides, and glucuronides), with the sugar moiety linked at position 3 [17,31]. They can also be found in aglycone form (quercetin, kaempferol, myricetin, and isorhamnetin) in wines as a result of hydrolysis of the heterosides during the vinification process [20,32]. Dihydroflavonols or flavononols are polyphenols that protect against UV radiation and are localized both in the skins and pulp but not in the seeds of grapes [22,33]. Unlike flavonols, their heterocycle is characterized by the absence of a double bond [20]. Astilbin (dihydroquercetin-3-O-rhamnoside) and engeletin (dihydrokaempferol-3-O-rhamnoside) are the principal dihydroflavonols identified in the skins of white grapes [30], along with dihydromyricetin-3-O-rhamnoside, which was confirmed in wine [34].

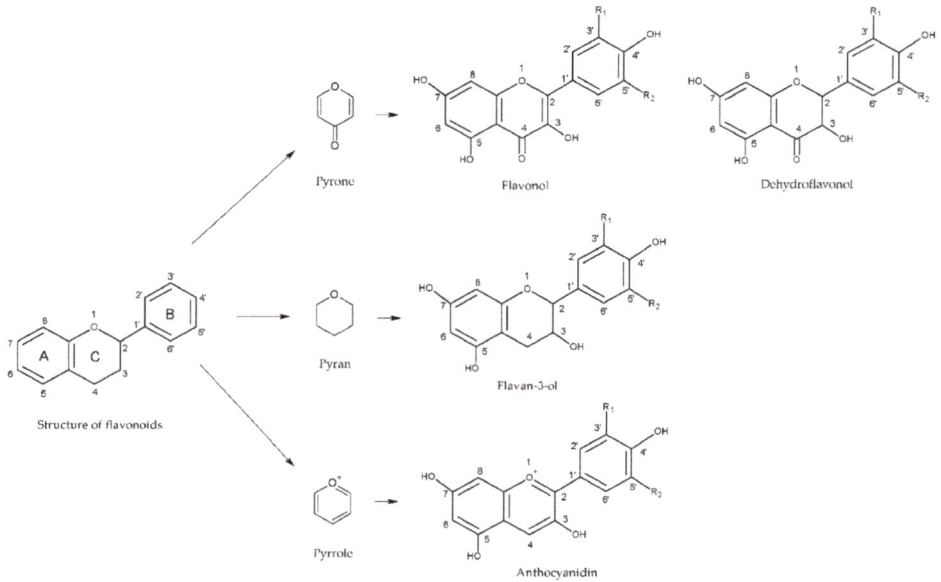

Figure 3. General flavonoid structure. Adapted from References [20,30]. R_1, R_2 = H, OH, OCH$_3$.

Flavanols, more accurately flavan-3-ols, contain a pyran heterocycle being hydroxylated in position 3 of the flavonoid skeleton [20]. They form a large group of catechins and their polymers (condensed tannins), which can be further categorized into procyanidins and delphinidins [20,24]. The catechin structure allows four isomeric monomers—(+/−) catechin and (+/−) epicatechin. If the gallate residue is in an isomeric *trans* position, we obtain four new isomers—(+/−) gallocatechin and (+/−) epigallocatechin (Figure 4). In addition, an important aspect is the potential galloylation (esterification of the hydroxyl group at position 3 with gallic acid), which considerably influences the astringency and bitterness of the resulting wines [20,30].

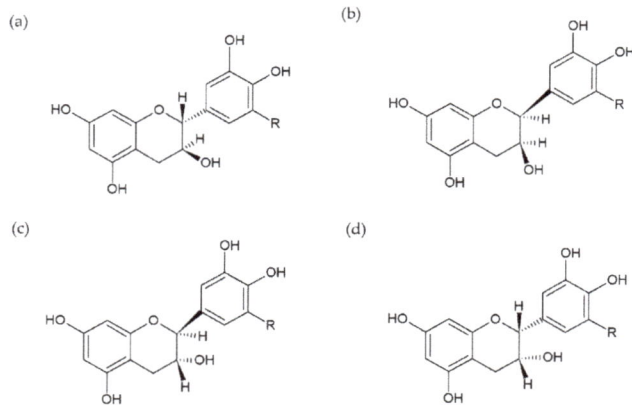

Figure 4. Diasteroisomers of catechin. Adapted from References [20,30]. (**a**) (+)-catechin (R = H); (+)-gallocatechin (R = OH); (**b**) (+)-epicatechin (R = H); (+)-epigallocatechin (R = OH); (**c**) (−)-catechin (R = H); (−)-gallocatechin (R = OH); (**d**) (−)-epicatechin (R = H); (−)-epigallocatechin (R = OH).

Flavanols may form oligomers and polymers, which are called condensed tannins or proanthocyanidins [24]. The term derives from their capacity to undertake hydrolysis under acidic conditions, which will cause the release of anthocyanidin pigments [20]. Furthermore, the intermediate unstable carbo-cations tend to combine with proteins or polysaccharides to produce stable complexes [35]. By coupling with the saliva proteins, tannins highlight specific astringency and bitterness sensations in red wines; at the same time, they participate in stabilizing the color as a result of coupling with anthocyanins through ethanol bridges [36]. Vivas patented a reactive proanthocyanidolic tannin [27] linked to an acetaldehyde molecule, which has the capacity to ensure color stabilization by linking to anthocyanins since the pre-fermentative stage and accelerate the polymerization reaction [19,37].

While condensed tannins are considered to be grape-derived, the tannins released by the wood during aging in barrels are called hydrolysable tannins or ellagitannins (ellagic acid is produced through their hydrolysis) [35].

Flavanol dimers, based on the bond between the flavanol subunits, are classified into as type-B procyanidins, which are dimers of two flavanol subunits linked either by a C4-C8 or a C4-C6 interflavan bond or type-A procyanidins, which have an ether bond between carbons C2-C5 or C2-C7 besides the interflavan bond [20,30].

Similarly, flavanols can form two classes of trimers—type-C procyanidins, linked by interflavan bonds; type-D procyanidins, linked by an interflavan bond and another flavano-ether bond (as in type-A procyanidins) [20,30].

In the case procyanidin trimers, up to 10 flavanol subunits are necessary (molecular mass of 600–3000 kDa), while the number of the flavanol subunits must be greater than 10 for condensed procyanidins (polymers), hence the denomination tannins or more correctly condensed tannins (molecular mass > 3000 kDa) [20,30].

Anthocyans, Anthocyanidins, and Anthocyanins are flavonoids that possess a pyrrole heterocycle and generate the characteristic reddish, bluish, and purple tints, as the main pigments in flowers and fruits [38,39]. Even though the general terminology is all-inclusive (comprising all three designations), anthocyanins represent the glycoside form and are made up of a sugar moiety linked to anthocyanidins (the aglycone form) through a glycosidic bond [22]. They are more stable in their glycosidic form than in the aglycone form. In red wines from the *Vitis vinifera* species, only monoglycosidic anthocyanins (C3 position) have been identified, with malvidin as the most abundant [22,40].

There is a link between the grape variety and the chromatic attributes of the wine; this is somewhat given by the present anthocyanidins and resulting compounds (through the reaction with glucose and its acylated forms), which vary widely depending on the grape variety [30,41]. This high chromatic variability among red varieties is responsible for the qualitative and quantitative differences in their composition, thus rendering them fit as traceability markers [20,41,42].

3. Polyphenol Fingerprinting and Analysis in Red Wines

3.1. Authenticity and Typicity Features

The authenticity and typicity aspects are quite important for all types of wine. While authenticity refers to the grape variety and the viticultural area of the resulting wine, typicity implies the analytical and sensorial characteristics; these are determined by the agro-biological particularities of the variety, by the agronomic and pedo-climatic conditions of the area of origin, as well as by the winemaking technology [43,44]. It is not always the case that authentic wines do necessarily possess typicity attributes. Depending on the vintage and the production method, two authentic wines may express a more or less pronounced typicity [43,45].

Grape must and wine polyphenols are correlated with various oenological and sensorial characteristics [46]. Up until now, the methods used for integrating polyphenol data into determining wine authenticity was mainly based on the total polyphenol content (TPC) [43]. It does neither reveal the chromatic aspects given by the anthocyanins, nor the amount of tannins that are very important in color stabilization [15,47]. As it does not offer precious information with regard to the pursued aim, the information on TPC should be regarded as complementary [43].

In the case of red wines, the authenticity and typicity characters may be achieved by the use of polyphenol fingerprinting, which is able to confirm the variety and geographical origin, besides the characteristic analyses specific to soil [44,45].

3.2. Polyphenols as Wine Traceability Markers

The analytical strategies, reported to date as being suitable for traceability, depend on several factors such as the nature of the compounds of interest, reliability, selectivity, accuracy, and reproducibility of the involved instrumentation [10]. In order to track a product and establish proper links to its history, the chosen analytical strategy must be accompanied by suitable analytical markers [45]. The term is generally referred to as a compound (chemical) pertaining to specific characteristics, which enable it useful (of interest for analytical purposes) for determination [10,48].

The requirements that an analytical marker has to possess in order to be deemed fit for food analysis are quite strict—stability during sample processing and storage; clear determination; adequate availability in the matrix; reliable identification; accessible reference substance [10,49].

Based on the vast diversity of polyphenol compounds in red grape varieties along with the structural transformations during wine elaboration and maturation, the employed qualitative and quantitative methods establish the resulting analytical fingerprint [10]. In this regard, to address the fingerprint and quantitative analysis, a "multivariate chromatographic fingerprint" would be necessary [50]. There are several reports showing the use of polyphenol classes as markers for analytical fingerprinting [6,49].

In particular, grape and wine quality, as well as varieties and cultivars are distinguished by determining the composition of phenolics [51,52]. Nevertheless, as described by Bertacchini, phenolics belong to the secondary or in-direct traceability indicators [9] as they cannot be linked to the same determinations in soil samples [9]. On the other hand, phenolics may be regarded as a fingerprint and can be used in authentication applications such as general characterization, discrimination based on geographic origin and variety, adulteration and/or contamination [9,45].

3.3. Profiling Applications

In this section, a collection of examples will be discussed dealing with the use of polyphenol profiling for red wines analysis according to the previously described sources of variability. In addition, an overview of the assessment of variety, vintage and geographical origin will be given, with respect to post-harvest treatments, processing and storage.

For fingerprint and profiling analyses, chemometrics has been regarded as the most versatile as it allows a complex approach to analyzing various known and unknown samples [9,53]. Accordingly, compositional profiles using polyphenols have been widely exploited [53,54].

As mentioned previously, the pattern of polyphenols in wine show pronounced variations and depend mainly on the grape cultivars, the corresponding growing conditions and the processing technology [44].

Among polyphenols, anthocyanins profile analysis supplies indications useful for authentication and differentiation of red wines by grape variety [15,44,45,55]. For instance, Geana aimed at providing an accurate classification of five Romanian red wines based on variety and vintage (six harvest years), using a linear discriminant analysis [56]. Besides the organic acids profile, NMR fingerprints and isotope analysis, they managed to assign vintage membership with a percentage of 91.64% using the anthocyanins and 92.56% using anthocyanin ratios, whereas variety discrimination resulted in 95.78% using the anthocyanins and 87.82% using anthocyanin ratios [1]. Similar results on anthocyanin composition and varietal classification were also reported in Czech wines [57].

It has been reported that during grape processing and wine storage anthocyanins that contain acyl and coumaryl moieties are the most stable. For this reason, these categories of anthocyanins have been proposed for red wine polyphenol fingerprinting [15]. More exactly, it should take into consideration both the sum of acylated and coumaroylated anthocyanins, as well as the ratio between their concentrations [15]. In addition, the relative proportion of acylated and non-acylated anthocyanins is characteristic of each cultivar [15,58]. Variety-based authentication using anthocyanin fingerprinting can also be correlated with the evaluation of shikimic acid content [59], although this is specific for white wines [60]; shikimic acid is found in low amounts in different fruit, including grapes. In red wines, its concentrations vary widely between 10–150 mg/L [43].

Anthocyanin profiles have also been effectively used for vintage discrimination [55,61,62]. In this case, one limitation of their use is that their original concentration changes (degradation reactions) during wine aging, thus affecting the resulting distribution [55,63].

In colder areas (positioned at the limit of vine cultivation), fingerprinting of red wines is based on the value of the ratio between the total polyphenol content and the content of malic acid; in fact, there is a stable inversely proportional correlation between the content of malic acid and anthocyanin concentration [43]. Moreover, it is mandatory that fingerprinting takes into consideration the sugar/titratable acidity ratio, which offers useful information on the technological maturity of the studied crop [43]. Vilanova also reported the significant influence of pH, titratable acidity, malic acid, and sugar/titratable acidity ratio on vintage classification, expressing their involvement in grape ripening [64]. In addition, as a consequence of climatic factors [65], flavanols, phenolic acids, and resveratrol supplied valuable information to the vintage classification [64].

In order to increase the level of objectivity associated with the polyphenol fingerprinting of red varieties, it was recently proposed the monitoring of some analytical parameters that characterize the berry skins of red grapes at technological maturity [17,22]; these parameters are responsible for the TPC in must and the resulting wines [66]. Under these circumstances, the focus is kept on the high concentration of anthocyanins and easily extractable tannins that are contained in grape skins, in comparison with the limited amounts of extractable tannins and lack of anthocyanins in the seeds [17,22]. Therefore, the relationship between skin and seed tannins and anthocyanins is regarded as an adequate analytical variable for determining the optimum harvest time [22,47].

In an attempt to classify red wines according to geographical origin, a collection of studies has reported the involvement of the polyphenolic profile in food authentication [44,67–70]. For example, flavanols, flavonols, and trans-resveratrol patterns were used for the classification of samples according to the wine production area [60,71,72]. Besides flavanols and flavonols, phenolic acids (gallic acid, *p*-coumaric acid, ferulic acid, and caffeic acid) were also successfully exploited to certify the production authenticity of red wines according to their regional identity [44,60,69,73–75]. Pisano noted that three malvidin-derived anthocyanins contributed mainly to the geographical origin discrimination of the studied samples [76]. In addition to the use of monomeric anthocyanins, Quaglieri indicated that wine samples from Rioja (Spain) and Aquitaine (France) were properly discriminated using supplementary markers such as dimers of flavan-3-ols, mean degree of polymerization (mDP), and polymerized pigments [77]. The authors also suggested a general tendency for a decrease in anthocyanin and tannin contents in older wines [77].

Biochemical features have previously been outlined in relation to metabolic changes across grape development with regard to varietal differences [78,79]. For example, using a series of chemical parameters related to the phenylpropanoid pathway such as the (+)-catechin/(−)epicatechin and malvidin-3-acetylglucoside/malvidin-3-coumaroylglucoside ratios, Muccillo observed a clear varietal classification of the analyzed wine samples [52]. A more comprehensive study enabled an inter-platform comparison of not-targeted metabolomics using polyphenol profiles for a protected denomination of origin (PDO) classification [80]. The authors also pointed out a strong relationship between the distribution of the polyphenol compounds and the biosynthetic pathway [80]. Further studies and reports regarding category attribution are compiled in Table 1, employing the use of a principal component analysis (PCA), a least significant difference (LSD) test or a linear discriminant analysis (LDA), in order to differentiate among the means of the variables (identified polyphenols).

Table 1. Examples of the use of polyphenol profiles for wine classification.

Application	Multivariate Approach	Source
Variety	PCA	[52,57,77,79,81–84]
	Discriminant analysis	[44,56,60,69,81,85]
Vintage	PCA	[73,86–88]
	Discriminant analysis	[56,69,87]
Geographical origin	PCA	[60,67,70,75,77,80,89,90]
	Discriminant analysis	[60,69,70,80]

4. Factors Affecting Polyphenol Concentration and Distribution in Wine

This section briefly addresses the agro-biologic aspects that must be considered in order to appropriately ascertain red wine polyphenol fingerprinting.

Typically, the amount and distribution of polyphenols in grapes and wine, similar to the diversity in the plant kingdom [6,10], are influenced by genetic and environmental factors [10]. Grape variety should be considered a determinant element in polyphenol fingerprinting of the resulting wines [15]. In turn, each variety highlights its attributes according to the soil composition, to the climatic conditions (hydro-thermic regimen) in the active vegetation period, as well as to the geographic positioning (latitude at which the vineyard is situated), and to the associated orographic factors (landscape—terrain, altitude, and exposition and incline of the terrain) [91,92].

Genetic factors. Clone selection may be perceived as an important biological instrument in improving grape quality, especially for polyphenol composition [93]. The genetic character of the variety is expressed mainly through the quantitative differences among polyphenol contents, even though the qualitative contrasts are highly important [16]. In this context, Gómez-Plaza identified remarkable seedlings from a collection of 143 intraspecific hybrids (Monastrell × Cabernet Sauvignon), by analyzing their anthocyanin profiles [94]. Similarly, based on the proanthocyanins in the skins and seeds of Monastrell and Syrah crosses, Hernández-Jiménez reported both qualitative and quantitative similarities between the clones and the origin varieties [95].

Ruiz-García and Gómez-Plaza pointed out the potential use of clone selection as a discriminant factor, by observing a significant differentiation of the clones' polyphenolic compositions [96]. In line with this, the anthocyanins, flavonols and hidroxycinnamic acids were successfully used for differentiating *Vitis vinifera* L. cv. 'Barbera' clones [21].

Pedo-climatic implications. The soil seems to be the least involved factor in the polyphenol fingerprinting of red wines. It tends to be expressed indirectly through its water and heat retaining capacities but also through its nutritional characteristics [97]. For example, the color and the textural composition of the soil affect heat absorption, having unfavorable implications on grape ripening and on plant protection against freezing [97]. Some authors reported the influence of soil on grape production per hectare, on sugar content, anthocyanins, tannins, and amino acids but also on carotenoids synthesis and their conversion to norisoprenoids [92,97,98]. Even though these findings are important, they do not highlight any link between the soil and polyphenol fingerprinting of the red wine obtained on the said soil.

The polyphenol composition of wines is also affected by the thermal and the hydric conditions [99]. Some authors suggest that in warm viticultural areas, the temperature may frequently reach levels that inhibit the formation of anthocyanins, thus affecting the chromatic characteristics of the grapes [100]. Other authors [101] showed that reducing the diurnal thermal fluctuations determines a simultaneous increase in grape maturation rates as well as the anthocyanin concentrations registered at harvest [101]. Besides the modification of the total anthocyanin content, a correlation between warm seasons and high concentrations of coumaroyl derivatives of malvidin, petunidin, and delphinidin have been reported [100]. Nevertheless, Tarara observed a link between high temperatures and a decrease in delphinidin, cyanidin, petunidin and peonidin concentrations in Merlot grapes exposed to sunlight but not in the case of malvindin derivatives, which were not affected [102].

Similarly, Nicholas observed a significant positive correlation between the content in anthocyanins and a thermal regime of 16–22 °C from ripening to the optimum harvest stage [103]; the same authors also reported a significant increase in the tannin content, explained by hight nocturnal temperatures before bud break and diurnal high temperatures from bud break until flowering [103]. Other studies noted that a moderate water deficit has a positive effect on red grape quality and implicitly on the quality of the resulting wines [92]. For example, the anthocyanin and tannin concentrations in grape skins are higher when the vineyard is exposed to a moderate water deficit [98,103,104].

The concentration and composition of polyphenols in the grapes vary with the harvest stage, besides the already mentioned variety specificity [40]. In this respect, the anthocyanins are accumulated in grape skins in the maturation stage, while flavanols are formed before ripening in higher amounts in seeds and skins and are lower in grape pulp [92]. Ollé reported a higher decrease rate for flavan-3-ol monomers in contrast to the polymers, thus associating the average level of polymerization with the progress of grape maturation [105]. Also, Verries observed a stationary concentration of seed and skin flavanols for up to one month before harvest [106].

In the years with unfavorable climate, wine producers are forced to prematurely harvest the grapes [92]. Under these conditions, the polyphenolic fingerprint of the resulting wine will not correspond to the real one (established under conditions of optimum polyphenol maturity) [92]. Moreover, a premature harvest also affects the wine's sensorial quality, deeming it unfit for the consumers' requirements [20,106–108].

Post-harvest processing and storage implications. During the maceration-fermentation stage, the correct management of temperature, as well as the appropriate duration of the contact between the solid and the liquid fractions of the must, is indispensable for the optimum extraction of polyphenols [43,109]. Usually, this stage begins with a temperature of 15–16 °C and increases by 1–2 °C per day up to the critical limit of 28 °C [41,43]. The tendency of decreasing the duration of the fermentation stage, along with the use of auxiliary enzymatic products or exogenous proanthocyanidolic tannins (from the seeds and skins of white grape varieties) may have both positive and negative impacts [41,110]. For instance, the addition of exogenous tannins results in the modification of the natural polyphenolic profile of the studied harvest, improving it. Therefore, when these practices are employed, the fingerprinting is limited to the polyphenols of the grapes harvested before applying any intervention [37,111,112]. On the other hand, a prolonged contact between the solid and the liquid fractions of the must allows for a better extraction of polyphenol compounds, especially catechins and proanthocyanidins [113,114] along with a sustained color stabilization [115].

Over time, several maceration-fermentation techniques have been developed for the production of red wines, with the main purpose of improving the polyphenol extraction [116]. Thermovinification, employed since the early 1970s [116,117], seeks the degradation of cell membranes of the grape berry hypoderm, thus facilitating anthocyanin extraction. In addition, the applied treatment also prevents browning by denaturing the oxidative enzymes like polyphenol oxidases [19,22,117,118]. Cold pre-fermentative maceration focuses on intensifying the extraction of polyphenols in aqueous media [119]. It takes place at temperatures lower than 10 °C and may vary from four to eight days [120,121] by favoring the selective extraction of low molecular mass anthocyanins and tannins [122].

During aging, some of the polyphenols from the wood barrels pass into the red wines, affecting their composition. Generally, aged wines are subjected to a slowly controlled oxidation, which gives rise to volatile compounds, along with polysaccharides and polyphenols that contribute to color stabilization [41,110]. The presence of ethanol acts against co-pigmentation, while acylated anthocyanins quickly disappear in several months after the end of fermentation [41,110]. Also, the amount of anthocyanins in red wines drops faster in the first years of bottle aging [123], reaching a minimum value of 0–50 mg/L [41]. Free anthocyanins may react with compounds containing diacetyl groups

(CH₃-CO-CO-CH₃), giving rise to castavinols that are not present in grapes but may spontaneously form in some red wines during aging in wood barrels [41,110]. These colorless compounds are capable of regenerating the chromatic characteristics of red wines to some extent as a result of the acidic environment, which allows the conversion of procyanidins into cyanidins [41,124,125].

The high degree of complexity of tannin-anthocyanin combinations has drawn the attention of numerous researchers since the late 1960s. Jurd showed that the flavylium ion (cation) could directly react with different components such as aminoacids, fluoroglucinol and catechin, producing a colorless flav-2-ene substituted at position C4 [126]. Somers suggested that this kind of reaction is involved in the process of wine aging [127]; the author observed the disappearance of anthocyanins, while the color remained stable or it intensified [127].

Chemical factors. Further research demonstrated that there are different mechanisms involved in the anthocyanin-tannin condensation reaction, which may result in compounds that carry various attributes, depending on the type of chemical bonds [22,41,86]. Thus, during the aging stage of red wines, three types of reactions have been identified and they will be briefly discussed [22,41,86].

Anthocyanins can act as cations on negatively charged carbon atoms in positions six and eight of the procyanidin ring, giving birth to a colorless compound called flavene. It requires the presence of oxygen or of an oxidant medium in order to recover its color [30]. One of the characteristics of procyanidins is that they form a carbo-cation after the addition of a proton to the molecule, which in turn may react with the nucleophilic sites (C6 and C8) of neutral anthocyanin molecules [22]. The resulting complex lacks color or may change it to red-orange in case of a water molecule loss [128]. This condensation reaction occurs in the absence of oxygen, it increases with the temperature and depends on the amount of anthocyanins in the wine [129]. In reality, the color of the resulting wine is modified according to the type of carbo-cation and also to the degree of anthocyanin polymerization [77,124,130].

During aging in contact with oak (chips, barrels), polyphenols (mainly catechins and proanthocyanidins) polymerize among them or with free anthocyanins. These new complex combinations stabilize and retain the reddish color of anthocyanins. Nonflavonoid tannins extracted from the wood of oak barrels do not appear to interfere in these condensation reactions but indirectly favor color stabilization by offering protection against oxidative degradation [131–133].

5. Concluding Remarks

In red grape varieties, the polyphenol composition can be perceived as the key element that significantly differentiates the final products, taking into account their specific agro-biological characteristics, in conjunction with the biosynthetic pathways of phenolics. Therefore, the harvest stage, the temperature, the duration of the maceration-fermentation process, and the vinification technique are among the factors involved in the quality of the resulting product. During winemaking, the technological conditions carry a notable weight on polyphenol fingerprinting, rendering the task of authenticating red wines one of great complexity. Subsequently, during storage, some modifications occur in red wines with regard to the interaction of certain phenolic compounds, which may influence the analyzed profiles. Collectively, the information presented in this review provides an overview of the challenges that arise in using polyphenol fingerprints as a tool for wine traceability and authenticity. Taking into account the overall impact of winemaking and storage conditions, as well as the information provided by the polyphenol synthesis pathways, the applicability of the current methods, should they be carefully employed, show promising potential in expediting and improving the verification/certification process for wine traceability.

Author Contributions: Writing—original draft preparation, L.M.P. and M.E.P.; supervision, M.E.P.

Funding: This research received no external funding.

Acknowledgments: This work is part of the Ph.D. thesis of L.M.P., supported by the Romanian Ministry of Education.

Beverages **2018**, *4*, 75

Conflicts of Interest: The authors declare no conflicts of interest. The founding sponsors had no role in the design of the study; in the collection, analyses, or interpretation of data; in the writing of the manuscript, and in the decision to publish the results.

References

1. Geana, E.I.; Popescu, R.; Costinel, D.; Dinca, O.R.; Stefanescu, I.; Ionete, R.E.; Bala, C. Verifying the red wines adulteration through isotopic and chromatographic investigations coupled with multivariate statistic interpretation of the data. *Food Control* **2016**, *62*, 1–9. [CrossRef]
2. Moe, T. Perspectives on traceability in food manufacture. *Trends Food Sci. Technol.* **1998**, *9*, 211–214. [CrossRef]
3. Regattieri, A.; Gamberi, M.; Manzini, R. Traceability of food products: General framework and experimental evidence. *J. Food Eng.* **2007**, *81*, 347–356. [CrossRef]
4. Bosona, T.; Gebresenbet, G. Food traceability as an integral part of logistics management in food and agricultural supply chain. *Food Control* **2013**, *33*, 32–48. [CrossRef]
5. Palade, M.; Popa, M.-E. Wine Traceability and Authenticity—A Literature Review. *Sci. Bull. Ser. F. Biotechnol.* **2014**, *XVIII*, 226–233.
6. Donno, D.; Boggia, R.; Zunin, P.; Cerutti, A.K.; Guido, M.; Mellano, M.G.; Prgomet, Z.; Beccaro, G.L. Phytochemical fingerprint and chemometrics for natural food preparation pattern recognition: an innovative technique in food supplement quality control. *J. Food Sci. Technol.* **2016**, *53*, 1071–1083. [CrossRef] [PubMed]
7. van Leeuwen, C.; Seguin, G. The concept of terroir in viticulture. *J. Wine Res.* **2006**, *17*, 1–10. [CrossRef]
8. Pavloušek, P.; Kumšta, M. Authentication of riesling wines from the Czech Republic on the basis of the non-flavonoid phenolic compounds. *Czech J. Food Sci.* **2013**, *31*, 474–482. [CrossRef]
9. Bertacchini, L.; Cocchi, M.; Li Vigni, M.; Marchetti, A.; Salvatore, E.; Sighinolfi, S.; Silvestri, M.; Durante, C. The Impact of Chemometrics on Food Traceability. In *Chemometrics in Food Chemistry*; Marini, F., Ed.; Elsevier: San Diego, CA, USA, 2013; Volume 28, pp. 371–410, ISBN 0922-3487.
10. Siracusa, L.; Ruberto, G. Plant Polyphenol Profiles as a Tool for Traceability and Valuable Support to Biodiversity. In *Polyphenols in Plants: Isolation, Purification and Extract Preparation*; Watson, R.R., Ed.; Academic Press, Elsevier: San Diego, CA, USA, 2014; pp. 15–33, ISBN 9780123979346.
11. Schlesier, K.; Fauhl-Hassek, C.; Forina, M.; Cotea, V.; Kocsi, E.; Schoula, R.; Jaarsveld, F.; Wittkowski, R. Characterisation and determination of the geographical origin of wines. Part I: overview. *Eur. Food Res. Technol.* **2009**, *230*, 1–13. [CrossRef]
12. Smeyers-Verbeke, J.; Jäger, H.; Lanteri, S.; Brereton, P.; Jamin, E.; Fauhl-Hassek, C.; Forina, M.; Römisch, U. Characterization and determination of the geographical origin of wines. Part II: descriptive and inductive univariate statistics. *Eur. Food Res. Technol.* **2009**, *230*, 15. [CrossRef]
13. Römisch, U.; Jäger, H.; Capron, X.; Lanteri, S.; Forina, M.; Smeyers-Verbeke, J. Characterization and determination of the geographical origin of wines. Part III: multivariate discrimination and classification methods. *Eur. Food Res. Technol.* **2009**, *230*, 31. [CrossRef]
14. Riccardo, F.; Annarita, P. Mass spectrometry in grape and wine chemistry. Part II: The consumer protection. *Mass Spectrom. Rev.* **2006**, *25*, 741–774. [CrossRef]
15. von Baer, D.; Rentzsch, M.; Hitschfeld, M.A.; Mardones, C.; Vergara, C.; Winterhalter, P. Relevance of chromatographic efficiency in varietal authenticity verification of red wines based on their anthocyanin profiles: Interference of pyranoanthocyanins formed during wine ageing. *Anal. Chim. Acta* **2008**, *621*, 52–56. [CrossRef] [PubMed]
16. Jackson, R. Vineyard practice. In *Wine Science*, 4th ed.; Academic Press: San Diego, CA, USA, 2014; pp. 143–306, ISBN 9780123814685.
17. Flamini, R.; Mattivi, F.; De Rosso, M.; Arapitsas, P.; Bavaresco, L. Advanced knowledge of three important classes of grape phenolics: Anthocyanins, stilbenes and flavonols. *Int. J. Mol. Sci.* **2013**, *14*, 19651–19669. [CrossRef] [PubMed]
18. Tsao, R. Chemistry and biochemistry of dietary polyphenols. *Nutrients* **2010**, *2*, 1231–1246. [CrossRef] [PubMed]
19. Vivas, N. *Les Composés Phénoliques et l'élaboration des Vins Rouges*; Éditions Féret: Bordeaux, France, 2007; ISBN 978-2-35156-004-4.

20. Moreno, J.; Peinado, R. Polyphenols. In *Enological Chemistry*; Elsevier Inc.: San Diego, CA, USA, 2012; pp. 53–76, ISBN 9780123884381.

21. Teixeira, A.; Eiras-Dias, J.; Castellarin, S.D.; Gerós, H. Berry phenolics of grapevine under challenging environments. *Int. J. Mol. Sci.* **2013**, *14*, 18711–18739. [CrossRef] [PubMed]

22. Jackson, R. Chemical Constituents of Grapes and Wine. In *Wine Science*, 4th ed.; Academic Press: San Diego, CA, USA, 2014; pp. 347–426, ISBN 9780123814685.

23. Knaggs, A.R. The biosynthesis of shikimate metabolites. *Nat. Prod. Rep.* **2001**, *18*, 334–355. [CrossRef] [PubMed]

24. Tsao, R.; McCallum, J. Chemistry of Flavonoids. In *Fruit and Vegetable Phytochemicals*; Wiley-Blackwell: Hoboken, NJ, USA, 2009; pp. 131–153, ISBN 9780813809397.

25. Kougan, G.B.; Tabopda, T.; Kuete, V.; Verpoorte, R. Simple Phenols, Phenolic Acids, and Related Esters from the Medicinal Plants of Africa. In *Medicinal Plant Research in Africa. Pharmacology and Chemistry*; Kuete, V., Ed.; Elsevier Inc.: San Diego, CA, USA, 2013; pp. 225–250, ISBN 9780124059276.

26. Gatto, P.; Vrhovsek, U.; Muth, J.; Segala, C.; Romualdi, C.; Fontana, P.; Pruefer, D.; Stefanini, M.; Moser, C.; Mattivi, F.; Velasco, R. Ripening and Genotype Control Stilbene Accumulation in Healthy Grapes. *J. Agric. Food Chem.* **2008**, *56*, 11773–11785. [CrossRef] [PubMed]

27. Vivas, N.; de Gaulejac, N.V.; Nonier, M.F. Estimation and quantification of wine phenolic compounds. *Bull. L'O.I.V* **2003**, *76*, 281–303. [CrossRef]

28. Temsamani, H.; Krisa, S.; Mérillon, J.-M.; Richard, T. Promising neuroprotective effects of oligostilbenes. *Nutr. Aging* **2015**, *3*, 49–54. [CrossRef]

29. Flamini, R.; Zanzotto, A.; de Rosso, M.; Lucchetta, G.; Vedova, A.D.; Bavaresco, L. Stilbene oligomer phytoalexins in grape as a response to Aspergillus carbonarius infection. *Physiol. Mol. Plant Pathol.* **2016**, *93*, 112–118. [CrossRef]

30. Terrier, N.; Poncet-Legrand, C.; Cheynier, V. Flavanols, Flavonols and Dihydroflavonols. In *Wine Chemistry and Biochemistry*; Moreno-Arribas, M.V., Polo, M.C., Eds.; Springer: New York, NY, USA, 2009; pp. 463–507, ISBN 9780387741161.

31. Castillo-Muñoz, N.; Gómez-Alonso, S.; García-Romero, E.; Gómez, M.V.; Velders, A.H.; Hermosín-Gutiérrez, I. Flavonol 3-O-glycosides series of Vitis vinifera cv. Petit Verdot red wine grapes. *J. Agric. Food Chem.* **2008**, *57*, 209–219. [CrossRef] [PubMed]

32. Georgiev, V.; Ananga, A.; Tsolova, V. Recent advances and uses of grape flavonoids as nutraceuticals. *Nutrients* **2014**, *6*, 391–415. [CrossRef] [PubMed]

33. Zhu, L.; Zhang, Y.; Lu, J. Phenolic contents and compositions in skins of red wine grape cultivars among various genetic backgrounds and originations. *Int. J. Mol. Sci.* **2012**, *13*, 3492–3510. [CrossRef] [PubMed]

34. Vitrac, X.; Castagnino, C.; Waffo-Téguo, P.; Delaunay, J.C.; Vercauteren, J.; Monti, J.P.; Deffieux, G.; Mérillon, J.M. Polyphenols newly extracted in red wine from Southwestern France by centrifugal partition chromatography. *J. Agric. Food Chem.* **2001**, *49*, 5934–5938. [CrossRef] [PubMed]

35. Niculescu, V.-C.; Paun, N.; Ionete, R.-E. The Evolution of Polyphenols from Grapes to Wines. *Grapes Wines-Adv. Prod. Process. Anal. Valorization* **2018**. [CrossRef]

36. Bosso, A.; Guaita, M. Study of some factors involved in ethanal production during alcoholic fermentation. *Eur. Food Res. Technol.* **2008**, *227*, 911–917. [CrossRef]

37. Croitoru, C.; Vivas, N.; Canariov, A.; Deaconu, L.; Codresi, C.; Hortolomei, G. Incidence of the treatment with oenological tannins on the red wines' sensorial profile. *Ann. Univ. Dunarea Jos Galati, Fascicle VI—Food Technol.* **2009**, *3*, 50–56.

38. He, F.; Mu, L.; Yan, G.L.; Liang, N.N.; Pan, Q.H.; Wang, J.; Reeves, M.J.; Duan, C.Q. Biosynthesis of anthocyanins and their regulation in colored grapes. *Molecules* **2010**, *15*, 9057–9091. [CrossRef] [PubMed]

39. Jaakola, L. New insights into the regulation of anthocyanin biosynthesis in fruits. *Trends Plant Sci.* **2013**, *18*, 477–483. [CrossRef] [PubMed]

40. Costa, E.; Cosme, F.; Jordão, A.M.; Mendes-Faia, A. Anthocyanin profile and antioxidant activity from 24 grape varieties cultivated in two Portuguese wine regions. *J. Int. des Sci. la Vigne du Vin* **2014**, *48*, 51–62. [CrossRef]

41. Ribereau-Gayon, P.; Glories, Y.; Maujean, A.; Dubourdieu, D. *Handbook of Enology Volume 2 The Chemistry of Wine Stabilization and Treatments*, 2nd ed.; John Wiley & Sons Ltd.: Hoboken, NJ, USA, 2006; ISBN 9780470010372.

42. Leicach, S.R.; Chludil, H.D. Plant Secondary Metabolites: Structure—Activity Relationships in Human Health Prevention and Treatment of Common Diseases. In *Studies in Natural Products Chemistry*; Elsevier: San Diego, CA, USA, 2014; Volume 42, pp. 267–304, ISBN 9780444632814.

43. Croitoru, C. *Oenologie. Inovari si noutati.*; Editura AGIR: Bucuresti, Romania, 2012; ISBN 978-973-720-463-9.

44. Di Paola-Naranjo, R.D.; Baroni, M.V.; Podio, N.S.; Rubinstein, H.R.; Fabani, M.P.; Badini, R.G.; Inga, M.; Ostera, H.A.; Cagnoni, M.; Gallegos, E.; et al. Fingerprints for main varieties of argentinean wines: Terroir differentiation by inorganic, organic, and stable isotopic analyses coupled to chemometrics. *J. Agric. Food Chem.* **2011**, *59*, 7854–7865. [CrossRef] [PubMed]

45. Versari, A.; Laurie, V.F.; Ricci, A.; Laghi, L.; Parpinello, G.P. Progress in authentication, typification and traceability of grapes and wines by chemometric approaches. *Food Res. Int.* **2014**, *60*, 2–18. [CrossRef]

46. Bakker, J.; Clarke, R.J. *Wine Flavour Chemistry*, 2nd ed.; Wiley-Blackwell: Hoboken, NJ, USA, 2012; ISBN 9781444330427.

47. Moreno, J.; Peinado, R. The Relationship Between Must Composition and Quality. In *Enological Chemistry*; Elsevier Inc.: San Diego, CA, USA, 2012; pp. 137–156, ISBN 9780123884381.

48. Amargianitaki, M.; Spyros, A. NMR-based metabolomics in wine quality control and authentication. *Chem. Biol. Technol. Agric.* **2017**, *4*, 1–12. [CrossRef]

49. Hakimzadeh, N.; Parastar, H.; Fattahi, M. Combination of multivariate curve resolution and multivariate classification techniques for comprehensive high-performance liquid chromatography-diode array absorbance detection fingerprints analysis of Salvia reuterana extracts. *J. Chromatogr. A* **2014**, *1326*, 63–72. [CrossRef] [PubMed]

50. Gad, H.A.; El-Ahmady, S.H.; Abou-Shoer, M.I.; Al-Azizi, M.M. Application of chemometrics in authentication of herbal medicines: A review. *Phytochem. Anal.* **2013**, *24*, 1–24. [CrossRef] [PubMed]

51. Guerrero, R.F.; Liazid, A.; Palma, M.; Puertas, B.; González-Barrio, R.; Gil-Izquierdo, Á.; García-Barroso, C.; Cantos-Villar, E. Phenolic characterisation of red grapes autochthonous to Andalusia. *Food Chem.* **2009**, *112*, 949–955. [CrossRef]

52. Muccillo, L.; Gambuti, A.; Frusciante, L.; Iorizzo, M.; Moio, L.; Raieta, K.; Rinaldi, A.; Colantuoni, V.; Aversano, R. Biochemical features of native red wines and genetic diversity of the corresponding grape varieties from Campania region. *Food Chem.* **2014**, *143*, 506–513. [CrossRef] [PubMed]

53. Saurina, J. Characterization of wines using compositional profiles and chemometrics. *TrAC Trends Anal. Chem.* **2010**, *29*, 234–245. [CrossRef]

54. Rubert, J.; Lacina, O.; Fauhl-Hassek, C.; Hajslova, J. Metabolic fingerprinting based on high-resolution tandem mass spectrometry: A reliable tool for wine authentication? *Anal. Bioanal. Chem.* **2014**, *406*, 6791–6803. [CrossRef] [PubMed]

55. Villano, C.; Lisanti, M.T.; Gambuti, A.; Vecchio, R.; Moio, L.; Frusciante, L.; Aversano, R.; Carputo, D. Wine varietal authentication based on phenolics, volatiles and DNA markers: State of the art, perspectives and drawbacks. *Food Control* **2017**, *80*, 1–10. [CrossRef]

56. Geana, E.I.; Popescu, R.; Costinel, D.; Dinca, O.R.; Ionete, R.E.; Stefanescu, I.; Artem, V.; Bala, C. Classification of red wines using suitable markers coupled with multivariate statistic analysis. *Food Chem.* **2016**, *192*, 1015–1024. [CrossRef] [PubMed]

57. Papoušková, B.; Bednář, P.; Hron, K.; Stávek, J.; Balík, J.; Myjavcová, R.; Barták, P.; Tománková, E.; Lemr, K. Advanced liquid chromatography/mass spectrometry profiling of anthocyanins in relation to set of red wine varieties certified in Czech Republic. *J. Chromatogr. A* **2011**, *1218*, 7581–7591. [CrossRef] [PubMed]

58. De Villiers, A.; Vanhoenacker, G.; Majek, P.; Sandra, P. Determination of anthocyanins in wine by direct injection liquid chromatography-diode array detection-mass spectrometry and classification of wines using discriminant analysis. *J. Chromatogr. A* **2004**, *1054*, 195–204. [CrossRef]

59. Mardones, C.; Hitschfeld, A.; Contreras, A.; Lepe, K.; Gutiérrez, L.; von Baer, D. Comparison of shikimic acid determination by capillary zone electrophoresis with direct and indirect detection with liquid chromatography for varietal differentiation of red wines. *J. Chromatogr. A* **2005**, *1085*, 285–292. [CrossRef] [PubMed]

60. Makris, D.P.; Kallithraka, S.; Mamalos, A. Differentiation of young red wines based on cultivar and geographical origin with application of chemometrics of principal polyphenolic constituents. *Talanta* **2006**, *70*, 1143–1152. [CrossRef] [PubMed]

61. Gonzalez-Neves, G.; Favre, G.; Gil, G.; Ferrer, M.; Charamelo, D. Effect of cold pre-fermentative maceration on the color and composition of young red wines cv. Tannat. *J. Food Sci. Technol.* **2015**, *52*, 3449–3457. [CrossRef] [PubMed]

62. Gustavo, G.-N.; Guzmán, F.; Diego, P.; Graciela, G. Anthocyanin profile of young red wines of Tannat, Syrah and Merlot made using maceration enzymes and cold soak. *Int. J. Food Sci. Technol.* **2015**, *51*, 260–267. [CrossRef]

63. He, F.; Liang, N.N.; Mu, L.; Pan, Q.H.; Wang, J.; Reeves, M.J.; Duan, C.Q. Anthocyanins and their variation in red wines I. Monomeric anthocyanins and their color expression. *Molecules* **2012**, *17*, 1571–1601. [CrossRef] [PubMed]

64. Vilanova, M.; Rodríguez, I.; Canosa, P.; Otero, I.; Gamero, E.; Moreno, D.; Talaverano, I.; Valdés, E. Variability in chemical composition of Vitis vinifera cv Mencía from different geographic areas and vintages in Ribeira Sacra (NW Spain). *Food Chem.* **2015**, *169*, 187–196. [CrossRef] [PubMed]

65. De La Presa-Owens, C.; Lamuela-Raventos, R.M.; Buxaderas, S.; De La Torre-Boronat, M.C. Characterization of Macabeo, Xarel.lo, and Parellada White Wines from the Penedès Region. II. *Am. J. Enol. Vitic.* **1995**, *46*, 529–541.

66. Chorti, E.; Kyraleou, M.; Kallithraka, S.; Pavlidis, M.; Koundouras, S.; Kanakis, I.; Kotseridis, Y. Irrigation and leaf removal effects on polyphenolic content of grapes and wines produced from cv. "Agiorgitiko" (Vitis vinifera L.). *Not. Bot. Horti Agrobot. Cluj-Napoca* **2016**, *44*, 133–139. [CrossRef]

67. Rodríguez-Delgado, M.-Á.; González-Hernández, G.; Conde-González, J.-E.; Pérez-Trujillo, J.-P. Principal component analysis of the polyphenol content in young red wines. *Food Chem.* **2002**, *78*, 523–532. [CrossRef]

68. Gómez-Ariza, J.L.; García-Barrera, T.; Lorenzo, F. Anthocyanins profile as fingerprint of wines using atmospheric pressure photoionisation coupled to quadrupole time-of-flight mass spectrometry. *Anal. Chim. Acta* **2006**, *570*, 101–108. [CrossRef]

69. Jaitz, L.; Siegl, K.; Eder, R.; Rak, G.; Abranko, L.; Koellensperger, G.; Hann, S. LC-MS/MS analysis of phenols for classification of red wine according to geographic origin, grape variety and vintage. *Food Chem.* **2010**, *122*, 366–372. [CrossRef]

70. Serrano-Lourido, D.; Saurina, J.; Hernández-Cassou, S.; Checa, A. Classification and characterisation of Spanish red wines according to their appellation of origin based on chromatographic profiles and chemometric data analysis. *Food Chem.* **2012**, *135*, 1425–1431. [CrossRef] [PubMed]

71. Rastija, V.; Srečnik, G. Polyphenolic composition of Croatian wines with different geographical origins. *Food Chem.* **2009**, *115*, 54–60. [CrossRef]

72. Radovanović, B.C.; Radovanović, A.N.; Souquet, J.-M. Phenolic profile and free radical-scavenging activity of Cabernet Sauvignon wines of different geographical origins from the Balkan region. *J. Sci. Food Agric.* **2010**, *90*, 2455–2461. [CrossRef] [PubMed]

73. Kallithraka, S.; Mamalos, A.; Makris, D.P. Differentiation of Young Red Wines Based on Chemometrics of Minor Polyphenolic Constituents. *J. Agric. Food Chem.* **2007**, *55*, 3233–3239. [CrossRef] [PubMed]

74. Kallithraka, S.; Tsoutsouras, E.; Tzourou, E.; Lanaridis, P. Principal phenolic compounds in Greek red wines. *Food Chem.* **2006**, *99*, 784–793. [CrossRef]

75. Salvatore, E.; Cocchi, M.; Marchetti, A.; Marini, F.; de Juan, A. Determination of phenolic compounds and authentication of PDO Lambrusco wines by HPLC-DAD and chemometric techniques. *Anal. Chim. Acta* **2013**, *761*, 34–45. [CrossRef] [PubMed]

76. Pisano, P.L.; Silva, M.F.; Olivieri, A.C. Anthocyanins as markers for the classification of Argentinean wines according to botanical and geographical origin. Chemometric modeling of liquid chromatography-mass spectrometry data. *Food Chem.* **2015**, *175*, 174–180. [CrossRef] [PubMed]

77. Quaglieri, C.; Prieto-Perea, N.; Berrueta, L.A.; Gallo, B.; Rasines-Perea, Z.; Jourdes, M.; Teissedre, P.L. Comparison of aquitaine and Rioja red wines: Characterization of their phenolic composition and evolution from 2000 to 2013. *Molecules* **2017**, *22*, 20. [CrossRef] [PubMed]

78. Degu, A.; Hochberg, U.; Sikron, N.; Venturini, L.; Buson, G.; Ghan, R.; Plaschkes, I.; Batushansky, A.; Chalifa-Caspi, V.; Mattivi, F.; Delledonne, M.; Pezzotti, M.; Rachmilevitch, S.; Cramer, G.R.; Fait, A. Metabolite and transcript profiling of berry skin during fruit development elucidates differential regulation between Cabernet Sauvignon and Shiraz cultivars at branching points in the polyphenol pathway. *BMC Plant Biol.* **2014**, *14*, 1–20. [CrossRef] [PubMed]

79. Figueiredo-González, M.; Martínez-Carballo, E.; Cancho-Grande, B.; Santiago, J.L.; Martínez, M.C.; Simal-Gándara, J. Pattern recognition of three Vitis vinifera L. red grapes varieties based on anthocyanin and flavonol profiles, with correlations between their biosynthesis pathways. *Food Chem.* **2012**, *130*, 9–19. [CrossRef]

80. Díaz, R.; Gallart-Ayala, H.; Sancho, J.V.; Nuñez, O.; Zamora, T.; Martins, C.P.B.; Hernández, F.; Hernández-Cassou, S.; Saurina, J.; Checa, A. Told through the wine: A liquid chromatography–mass spectrometry interplatform comparison reveals the influence of the global approach on the final annotated metabolites in non-targeted metabolomics. *J. Chromatogr. A* **2016**, *1433*, 90–97. [CrossRef] [PubMed]

81. Vaclavik, L.; Lacina, O.; Hajslova, J.; Zweigenbaum, J. The use of high performance liquid chromatography-quadrupole time-of-flight mass spectrometry coupled to advanced data mining and chemometric tools for discrimination and classification of red wines according to their variety. *Anal. Chim. Acta* **2011**, *685*, 45–51. [CrossRef] [PubMed]

82. Ma, Y.; Tanaka, N.; Vaniya, A.; Kind, T.; Fiehn, O. Ultrafast Polyphenol Metabolomics of Red Wines Using MicroLC-MS/MS. *J. Agric. Food Chem.* **2016**, *64*, 505–512. [CrossRef] [PubMed]

83. Heras-Roger, J.; Díaz-Romero, C.; Darias-Martín, J. A comprehensive study of red wine properties according to variety. *Food Chem.* **2016**, *196*, 1224–1231. [CrossRef] [PubMed]

84. Ivanova-Petropulos, V.; Ricci, A.; Nedelkovski, D.; Dimovska, V.; Parpinello, G.P.; Versari, A. Targeted analysis of bioactive phenolic compounds and antioxidant activity of Macedonian red wines. *Food Chem.* **2015**, *171*, 412–420. [CrossRef] [PubMed]

85. Sen, I.; Tokatli, F. Authenticity of wines made with economically important grape varieties grown in Anatolia by their phenolic profiles. *Food Control* **2014**, *46*, 446–454. [CrossRef]

86. Ivanova, V.; Vojnoski, B.; Stefova, M. Effect of winemaking treatment and wine aging on phenolic content in Vranec wines. *J. Food Sci. Technol.* **2012**, *49*, 161–172. [CrossRef] [PubMed]

87. Nikfardjam, M.S.P.; Márk, L.; Avar, P.; Figler, M.; Ohmacht, R. Polyphenols, anthocyanins, and trans-resveratrol in red wines from the Hungarian Villány region. *Food Chem.* **2006**, *98*, 453–462. [CrossRef]

88. García-Marino, M.; Hernández-Hierro, J.M.; Santos-Buelga, C.; Rivas-Gonzalo, J.C.; Escribano-Bailón, M.T. Multivariate analysis of the polyphenol composition of Tempranillo and Graciano red wines. *Talanta* **2011**, *85*, 2060–2066. [CrossRef] [PubMed]

89. Bellomarino, S.A.; Conlan, X.A.; Parker, R.M.; Barnett, N.W.; Adams, M.J. Geographical classification of some Australian wines by discriminant analysis using HPLC with UV and chemiluminescence detection. *Talanta* **2009**, *80*, 833–838. [CrossRef] [PubMed]

90. Radovanovic, A.; Jovancicevic, B.; Arsic, B.; Radovanovic, B.; Bukarica, L.G. Application of non-supervised pattern recognition techniques to classify Cabernet Sauvignon wines from the Balkan region based on individual phenolic compounds. *J. Food Compos. Anal.* **2016**, *49*, 42–48. [CrossRef]

91. Anesi, A.; Stocchero, M.; Dal Santo, S.; Commisso, M.; Zenoni, S.; Ceoldo, S.; Tornielli, G.B.; Siebert, T.E.; Herderich, M.; Pezzotti, M.; Guzzo, F. Towards a scientific interpretation of the terroir concept: plasticity of the grape berry metabolome. *BMC Plant Biol.* **2015**, *15*, 191. [CrossRef] [PubMed]

92. Jackson, R. Site Selection and Climate. In *Wine Science*, 4th ed.; Academic Press: San Diego, CA, USA, 2014; pp. 307–346, ISBN 9780123814685.

93. Revilla, E.; García-Beneytez, E.; Cabello, F. Anthocyanin fingerprint of clones of Tempranillo grapes and wines made with them. *Aust. J. Grape Wine Res.* **2009**, *15*, 70–78. [CrossRef]

94. Gómez-Plaza, E.; Gil-Muñoz, R.; Hernández-Jiménez, A.; López-Roca, J.M.; Ortega-Regules, A.; Martínez-Cutillas, A. Studies on the anthocyanin profile of Vitis Vinifera intraspecific hybrids (Monastrell × Cabernet Sauvignon). *Eur. Food Res. Technol.* **2008**, *227*, 479–484. [CrossRef]

95. Hernández-Jiménez, A.; Gómez-Plaza, E.; Martínez-Cutillas, A.; Kennedy, J.A. Grape Skin and Seed Proanthocyanidins from Monastrell × Syrah Grapes. *J. Agric. Food Chem.* **2009**, *57*, 10798–10803. [CrossRef] [PubMed]

96. Ruiz-García, Y.; Gómez-Plaza, E. Elicitors: A Tool for Improving Fruit Phenolic Content. *Agriculture* **2013**, *3*, 33–52. [CrossRef]

97. Deloire, A.; Vaudour, E.; Carey, V.; Bonnardot, V.; Leeuwen, C.V. Grapevine responses to terroir: A global approach. *J. Int. Sci. Vigne Vin* **2005**, *39*, 149–162. [CrossRef]

98.	Zsófi, Z.; Gál, L.; Szilágyi, Z.; Szűcs, E.; Marschall, M.; Nagy, Z.; Bálo, B. Use of stomatal conductance and pre-dawn water potential to classify terroir for the grape variety Kékfrankos. *Aust. J. Grape Wine Res.* **2009**, *15*, 36–47. [CrossRef]

99.	Mira de Orduña, R. Climate change associated effects on grape and wine quality and production. *Food Res. Int.* **2010**, *43*, 1844–1855. [CrossRef]

100.	Downey, M.O.; Dokoozlian, N.K.; Krstic, M.P. Cultural practice and environmental impacts on the flavonoid composition of grapes and wine: A review of recent research. *Am. J. Enol. Vitic.* **2006**, *57*, 257–268.

101.	Cohen, S.D.; Tarara, J.M.; Kennedy, J.A. Assessing the impact of temperature on grape phenolic metabolism. *Anal. Chim. Acta* **2008**, *621*, 57–67. [CrossRef] [PubMed]

102.	Tarara, J.M.; Lee, J.; Spayd, S.E.; Scagel, C.F. Berry Temperature and Solar Radiation Alter Acylation, Proportion, and Concentration of Anthocyanin in Merlot Grapes. *Am. J. Enol. Vitic.* **2008**, *59*, 235–247.

103.	Nicholas, K.A.; Matthews, M.A.; Lobell, D.B.; Willits, N.H.; Field, C.B. Effect of vineyard-scale climate variability on Pinot noir phenolic composition. *Agric. For. Meteorol.* **2011**, *151*, 1556–1567. [CrossRef]

104.	Castellarin, S.D.; Matthews, M.A.; Di Gaspero, G.; Gambetta, G.A. Water deficits accelerate ripening and induce changes in gene expression regulating flavonoid biosynthesis in grape berries. *Planta* **2007**, *227*, 101–112. [CrossRef] [PubMed]

105.	Ollé, D.; Guiraud, J.L.; Souquet, J.M.; Terrier, N.; Ageorges, A.; Cheynier, V.; Verries, C. Effect of pre- and post-veraison water deficit on proanthocyanidin and anthocyanin accumulation during Shiraz berry development. *Aust. J. Grape Wine Res.* **2011**, *17*, 90–100. [CrossRef]

106.	Verries, C.; Guiraud, J.-L.; Souquet, J.-M.; Vialet, S.; Terrier, N.; Ollé, D. Validation of an Extraction Method on Whole Pericarp of Grape Berry (Vitis vinifera L. cv. Shiraz) to Study Biochemical and Molecular Aspects of Flavan-3-ol Synthesis during Berry Development. *J. Agric. Food Chem.* **2008**, *56*, 5896–5904. [CrossRef] [PubMed]

107.	Cadot, Y.; Caillé, S.; Samson, A.; Barbeau, G.; Cheynier, V. Sensory representation of typicality of Cabernet franc wines related to phenolic composition: Impact of ripening stage and maceration time. *Anal. Chim. Acta* **2012**, *732*, 91–99. [CrossRef] [PubMed]

108.	Jackson, R. Grapevine Structure and Function. In *Wine Science*, 4th ed.; Academic Press: San Diego, CA, USA, 2014; pp. 69–141, ISBN 9780123814685.

109.	Croitoru, C. *Tratat de ştiinţă şi Inginerie Oenologică. Produse de Elaborare şi Maturare a Vinurilor*; Editura AGIR: Bucuresti, Romania, 2009; ISBN 978-973-720-065-5.

110.	Jackson, R. Post-Fermentation Treatments and Related Topics. In *Wine Science*, 4th ed.; Academic Press: San Diego, CA, USA, 2014; pp. 535–676, ISBN 9780123814685.

111.	Bautista-Ortín, A.B.; Cano-Lechuga, M.; Ruiz-García, Y.; Gómez-Plaza, E. Interactions between grape skin cell wall material and commercial enological tannins. Practical implications. *Food Chem.* **2014**, *152*, 558–565. [CrossRef] [PubMed]

112.	Malacarne, M.; Nardin, T.; Bertoldi, D.; Nicolini, G.; Larcher, R. Verifying the botanical authenticity of commercial tannins through sugars and simple phenols profiles. *Food Chem.* **2016**, *206*, 274–283. [CrossRef] [PubMed]

113.	Ivanova, V.; Dörnyei, Á.; Márk, L.; Vojnoski, B.; Stafilov, T.; Stefova, M.; Kilár, F. Polyphenolic content of Vranec wines produced by different vinification conditions. *Food Chem.* **2011**, *124*, 316–325. [CrossRef]

114.	Francesca, N.; Romano, R.; Sannino, C.; Le Grottaglie, L.; Settanni, L.; Moschetti, G. Evolution of microbiological and chemical parameters during red wine making with extended post-fermentation maceration. *Int. J. Food Microbiol.* **2014**, *171*, 84–93. [CrossRef] [PubMed]

115.	Xia, E.-Q.; Deng, G.-F.; Guo, Y.-J.; Li, H.-B. Biological activities of polyphenols from grapes. *Int. J. Mol. Sci.* **2010**, *11*, 622–646. [CrossRef] [PubMed]

116.	Sacchi, K.L.; Bisson, L.F.; Adams, D.O. A review of the effect of winemaking techniques on phenolic extraction in red wines. *Am. J. Enol. Vitic.* **2005**, *56*, 197–206.

117.	Wagener, G.W.W. The Effect of Different Thermovinification Systems on Red Wine Quality. *Am. J. Enol. Vitic.* **1981**, *32*, 179–184.

118.	Moreno, J.; Peinado, R. Redox Phenomena in Must and Wine. In *Enological Chemistry*; Elsevier Inc.: San Diego, CA, USA, 2012; pp. 289–302, ISBN 9780123884381.

119.	Gomez-Miguez, M.; Gonzalez-Miret, M.L.; Heredia, F.J. Evolution of colour and anthocyanin composition of Syrah wines elaborated with pre-fermentative cold maceration. *J. Food Eng.* **2007**, *79*, 271–278. [CrossRef]

120. Heredia, F.J.; Escudero-Gilete, M.L.; Hernanz, D.; Gordillo, B.; Meléndez-Martínez, A.J.; Vicario, I.M.; González-Miret, M.L. Influence of the refrigeration technique on the colour and phenolic composition of syrah red wines obtained by pre-fermentative cold maceration. *Food Chem.* **2010**, *118*, 377–383. [CrossRef]

121. Ortega-Heras, M.; Pérez-Magariño, S.; González-Sanjosé, M.L. Comparative study of the use of maceration enzymes and cold pre-fermentative maceration on phenolic and anthocyanic composition and colour of a Mencía red wine. *LWT—Food Sci. Technol.* **2012**, *48*, 1–8. [CrossRef]

122. Álvarez, I.; Aleixandre, J.L.; García, M.J.; Lizama, V. Impact of prefermentative maceration on the phenolic and volatile compounds in Monastrell red wines. *Anal. Chim. Acta* **2006**, *563*, 109–115. [CrossRef]

123. Czibulya, Z.; Kollár, L.; Nikfardjam, M.P.; Kunsági-Máté, S. The effect of temperature on the color of red wines. *J. Food Sci.* **2012**, *77*, C880–C885. [CrossRef] [PubMed]

124. Gambuti, A.; Rinaldi, A.; Ugliano, M.; Moio, L. Evolution of phenolic compounds and astringency during aging of red wine: Effect of oxygen exposure before and after bottling. *J. Agric. Food Chem.* **2013**, *61*, 1618–1627. [CrossRef] [PubMed]

125. Jackson, R. Fermentation. In *Wine Science*, 4th ed.; Academic Press: San Diego, CA, USA, 2014; pp. 427–534, ISBN 9780123814685.

126. Jurd, L. Anthocyanidins and related compounds—XI: Catechin-flavylium salt condensation reactions. *Tetrahedron* **1967**, *23*, 1057–1064. [CrossRef]

127. Somers, T.C. The polymeric nature of wine pigments. *Phytochemistry* **1971**, *10*, 2175–2186. [CrossRef]

128. Moreno, J.; Peinado, R. Aging. In *Enological Chemistry*; Elsevier Inc.: San Diego, CA, USA, 2012; pp. 389–403, ISBN 9780123884381.

129. Lorrain, B.; Ky, I.; Pechamat, L.; Teissedre, P.L. Evolution of analysis of polyhenols from grapes, wines, and extracts. *Molecules* **2013**, *18*, 1076–1100. [CrossRef] [PubMed]

130. Sun, B.; Neves, A.C.; Fernandes, T.A.; Fernandes, A.L.; Mateus, N.; De Freitas, V.; Leandro, C.; Spranger, M.I. Evolution of phenolic composition of red wine during vinification and storage and its contribution to wine sensory properties and antioxidant activity. *J. Agric. Food Chem.* **2011**, *59*, 6550–6557. [CrossRef] [PubMed]

131. Vivas, N.; Vivas de Gaulejac, N.; Nonier, M.F. Quelques aspects cinétiques de la consommation de l'oxygène et conséquences technologiques des oxydations dans les vins rouges: Partie 2/2: Discussion. *Rev. Oenol.* **2014**, *41*, 29–30.

132. Cozzolino, D. The role of visible and infrared spectroscopy combined with chemometrics to measure phenolic compounds in grape and wine samples. *Molecules* **2015**, *20*, 726–737. [CrossRef] [PubMed]

133. Dumitriu, G.-D.; de Lerma, N.L.; Cotea, V.V.; Zamfir, C.-I.; Peinado, R.A. Effect of aging time, dosage and toasting level of oak chips on the color parameters, phenolic compounds and antioxidant activity of red wines (var. Fetească neagră). *Eur. Food Res. Technol.* **2016**, *242*, 2171–2180. [CrossRef]

beverages

MDPI

Review

From the Field to the Bottle—An Integrated Strategy for Wine Authenticity

Leonor Pereira [1,2,†], **Sónia Gomes** [1,2,†], **Sara Barrias** [1,2,†], **Elisa Preto Gomes** [1,3],
Margarida Baleiras-Couto [2,4], **José Ramiro Fernandes** [5] and **Paula Martins-Lopes** [1,2,*]

[1]   School of Life Science and Environment, University of Trás-os-Montes and Alto Douro, 5000-901 Vila Real,
      Portugal; leopereira@utad.pt (L.P.); sgomes@utad.pt (S.G.); sarabarrias@hotmail.com (S.B.);
      mgomes@utad.pt (E.P.G.)
[2]   Faculty of Sciences, BioISI—Biosystems & Integrative Sciences Institute Campo Grande,
      University of Lisboa, C8 bdg, 1749-016 Lisboa, Portugal; margarida.couto@iniav.pt
[3]   Geosciences Center, Faculty of Sciences and Technology, University of Coimbra, 3030-790 Coimbra, Portugal
[4]   National Institute for Agricultural and Veterinary Research (INIAV), 2565-191 Dois Portos, Portugal
[5]   CQVR and Department of Physics, University of Trás-os-Montes & Alto Douro, 5001-801 Vila Real, Portugal;
      jraf@utad.pt
*   Correspondence: plopes@utad.pt
†   These authors contributed equally to this work.

Received: 3 August 2018; Accepted: 19 September 2018; Published: 1 October 2018

Abstract: The wine sector is one of the most economically important agro-food businesses. The wine market value is largely associated to terroir, in some cases resulting in highly expensive wines that attract fraudulent practices. The existent wine traceability system has some limitations that can be overcome with the development of new technological approaches that can tackle this problem with several means. This review aims to call attention to the problem and to present several strategies that can assure a more reliable and authentic wine system, identifying existent technologies developed for the sector, which can be incorporated into the current traceability system.

Keywords: wine authenticity; geographical origin; grapevine varietal identification and discrimination; bio-geochemical strategy

1. Wine Authenticity

For all food and beverage production, it is fundamental to employ procedures that control the quality, safety, and authenticity of products. Authenticity in the food industry, in particular in added-value food products, such as wine, has been a major concern that has challenged researchers to develop reliable and feasible technologies for such a purpose [1,2]. The wine sector is a billion-euro business, where highly quoted wines are the preferential target for fraudulent practices. Their quality is known to be deeply influenced by many factors, and amongst them the grapevine varieties used, origins, and growing conditions play a major role [3,4]. The quality of the final product is also strongly influenced by the physical, chemical, and molecular biological transformations involved in the process of winemaking. These transformations are a result of the action of various enzymes, mainly from yeasts, and specific bacteria, which are responsible for many fermentation steps occurring during winemaking [5]. These parameters, related to the history and provenance of wines, strongly set its commercial value; therefore, a set of rigorous legal guidelines and a strong organizational culture towards quality control are required to guarantee the safety and quality of wines [2].

Nowadays, both consumers and winemakers show an increasing interest in finding different ways to assess the authenticity of their products. In this direction, traceability systems can be used as a risk management tool, utilized to easily trace the origin and the overall vinification process [6].

In Europe, traceability systems are applied to promote and protect certain denominations, such as the Protected Designation of Origin (PDO), Protected Geographical Indication (PGI), and Traditional Specialties Guaranteed (TSG). These designations of origin are conferred to high quality agricultural products, like wine, which are strictly linked to their origin area and specific viticulture and oenological practices [7,8].

A specific case of geographical indication concerning wine is its terroir, a term that relates the origin of a certain wine to a very specific area and includes specific geologic and geomorphologic boundaries (e.g., soil, topography, climate, landscape characteristics, and biodiversity features). In vitivinicultural, terroir consists of an area, in which the interactions between the identifiable physical and biological environment and the vitivinicultural practices used will provide distinctive characteristics of the wine originated from that area in particular [9].

The International Organization of Vine and Wine (OIV) has set clearly the definitions of Recognized Geographical Indication and the Recognized Appellation of Origin [10]. Both have the name of the country, region, or place in the label, which requires previous recognition of the authorities of the country concerned, and consist of products of quality and/or characteristics linked to the geographical milieu (natural and human factors), requiring that grapes are harvested in the defined denomination. However, the Recognized Appellation of Origin entails that the products' characteristics are due exclusively or essentially to the geographical location and that the grape transformation is performed in the defined area.

The denominations of origin have been established for quite a long time in Europe, being the first regions defined in the 18th century: Chianti, an Italian denomination, was established in 1716; Tokaj, a Hungarian denomination, was established in 1757; and the Douro Wine Region, a Portuguese denomination, was created in 1756 as the first wine appellation in the world that had, apart from the definition of the wine producing area, regulations on producing methodologies and trade rules. Nowadays, the Appellations of Origin are spread throughout the world in all continents, making it necessary to develop tools that are more sensitive in sensing the differences among regions. Some of the Appellations of Origin are represented in Figure 1.

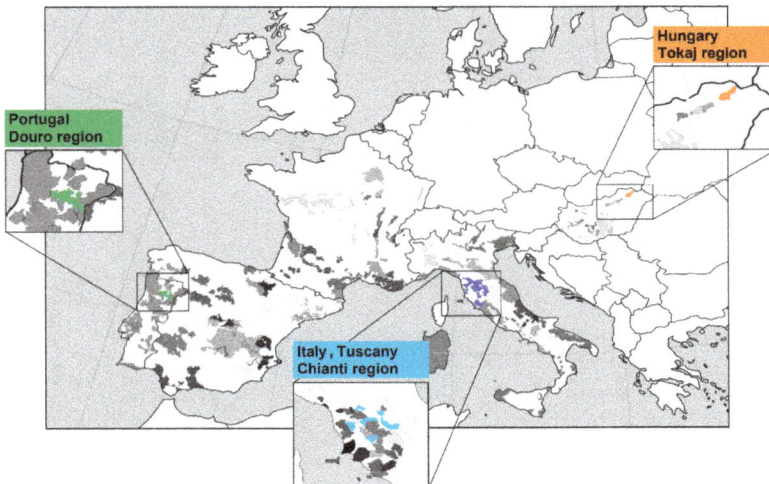

Figure 1. Appellations of Origin in the Old World. In grey are some of the Appellations of Origin in several wine producing areas and highlighted in color are the first Appellations (in blue the Chianti region; in green the Douro region; and in orange the Tokaj region). Adapted from Vineyards © [11] (accessed at 31 July 2018).

The main grapevine varieties from the old world have also been taken to new wine producing countries, giving rise to a diversity of wines commercially available in the market that have the same genetic origin. The most widely spread grapevine variety used for wine production is Cabernet Sauvignon, covering an area of 341,000 ha, mainly grown in China, France, Chile, the United States, Australia, Spain, Argentina, Italy, and South Africa [12]. Thus, the geographical origin identification is crucial under a well-established authenticity scheme.

The label present in wine bottles can also contain further information, such as the grapevine varietal composition. Nevertheless, as for geographical origin, the varietal composition is also regulated by the OIV. The label can state the variety if at least 75% of the grapes belong to such a grapevine variety and is listed in the denomination and if it attributes a specific characteristic to the wine [10]. Wines mentioning two varieties must comprehend exclusively these varieties and they should contain more than 15% of the listed varieties, which must be indicated by decreasing order of importance [10]. When more than two varieties are listed, the label must contain their respective percentages [10].

When production is carried out according to the standardized procedures, it normally results in final products with a high quality, which translates to higher prices at the sale point. Unfortunately, these financial benefits attract the production of counterfeit products and illegal food trades [13]. The dilution of wines with water, addition of alcohol or coloring and flavoring substances, blending with a wine of a lesser quality, and the mislabeling and misrepresentation of grapevine varieties and geographical origin are different kinds of frauds that can be referred to as examples of wine adulteration [14]. With the increasing occurrence of fraudulent practices, fast, reliable, and competent methods are needed to tackle authentication challenges and ensure products' quality. These strategies should guarantee the consumer's protection against mislabeling information of the purchased products, and the honest producers' defense from prejudicial competitors [1,15].

Two of the main requirements for the assessment of wine authenticity are the determination of its geographical origin, since the area of production is associated with the originality and quality of the characteristics of products, and the grapevine varietal identification, with compositional and sensory parameters being highly dependent on the variety (or varieties) used to produce a certain wine [3].

Several methods based on the analysis of metabolites, such as volatile compounds [16], amino acids and proteins [17,18], phenolic compounds [14], anthocyanins [19], mineral composition, and isotope identification [20], have been developed for the assessment of wine authenticity. These can be used for grapevine varietal identification, and some can also be used for geographical origin determination. Promising results have been obtained, however, the metabolic composition of grapes and wines is influenced not only by environmental conditions, cultural practices, and climate changes, but also by the production systems and processing methods used. On the other hand, these variables do not affect the grapevine genotype. Therefore, varietal identification and discrimination might be more accurate and efficient when DNA-based methodologies are used [21,22].

Regarding DNA methodologies, grapevine varietal identification is currently easily guaranteed with the use of simple sequence repeat (SSR) markers, approved and supported by the OIV [23]. Once developed, they are easy and inexpensive to use, and data can be readily compared among laboratories. Nevertheless, drawbacks have been reported for the application of these markers in wine samples due to the low amount of DNA isolated from this type of matrix [7]. The development of single nucleotide polymorphism (SNP) markers is also being considered as an alternative to SSRs. SNP markers have proven to be highly stable and repeatable, with a high discriminating power for grapevine varieties [24].

Although a large number of potential methods/technologies have been developed throughout the years that aim to target wine authenticity, the main system used is still mainly based on traceability systems. These traceability systems are mandatory by Reg. 178/2002 [25] for all agri-food products, including wine. However, the traceability systems are mainly based on registrations that can be adulterated, therefore, constituting a fragility of the system. Therefore, it is important that multidisciplinary strategies that can assist and control traceability systems are developed so the entire chain can be better protected.

2. The Importance of an Integrated Strategy

The need to develop multidisciplinary strategies in wine authenticity is the only reliable way of guaranteeing that all different terroir levels are contemplated in the analysis. As previously mentioned, the terroir is a result of multidimensional parameters, including soil, climate, and biodiversity features. The use of a unique technology capable of evaluating all these dimensions has so far not been accomplished. Nonetheless, several technologies have proven to be efficient to evaluate one of the components required to design the authenticity plan (Figure 2).

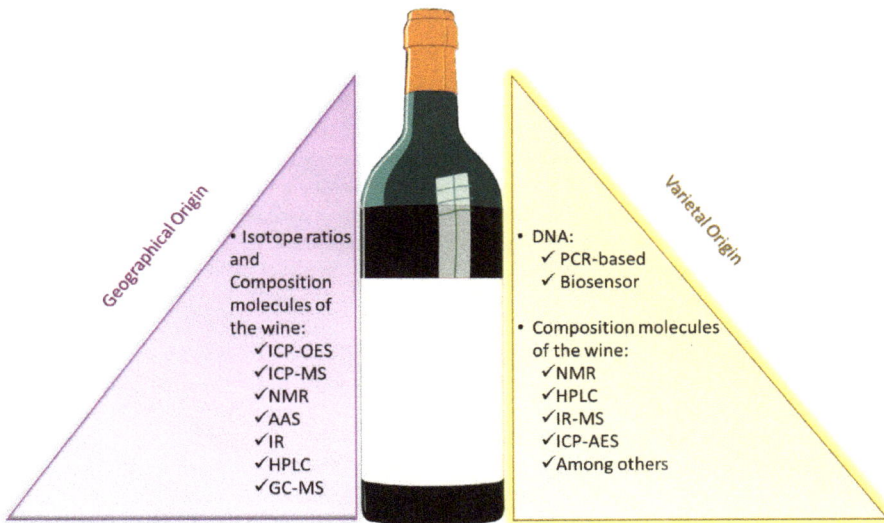

Figure 2. Technologies suitable to assess Geographical Origin and Varietal Identification.

The use of DNA-based detection systems has been extended to a wide variety of food products, including wine, as this remains the most reliable methods for varietal identification purposes. Among the different DNA-based systems, the use of biosensors has emerged as an attractive and alternative method for food authenticity. An optical biosensor system has been used for such a purpose, presenting several advantages, such as low cost, real-time measurement, and label free detection [26,27].

Even though varietal identification is possible when molecular markers are applied to wine samples, the grape origin cannot be detected by these means. Geographical origin can be achieved through chemical and isotopic techniques. Soil related fingerprinting plays a primary role in the determination of the geographical provenience since there is a direct correlation between the chemical composition of the wine and the soil composition, particularly the $^{87}Sr/^{86}Sr$ ratio, which can be used for the definition of the different denominations of origin [28]. Additionally, the mostly widely applied methods that intend to combine both botanical and geographical origin in wine samples are based on spectroscopic and/or spectrometric approaches [29]. These techniques are high-throughput approaches that are based on big datasets, with a previous collection of data considering several varieties and production years for each particular region associated with a statistical treatment. However, these approaches are not always efficient considering varietal identification [30].

Therefore, instead of aiming to define a unique technology, the integration of two-dimensional strategies, one for geographical origin determination and another for varietal identification and quantification, can be a more convenient and reliable way of tackling this issue. This type of approach has been already suggested by Fernandes et al. [31] as a bio-geochemical strategy, considering the grapevine composition through a biological method, and the definition of provenance based on geochemical determination.

3. Determination of the Region of Provenience

Wine is one of the main food products commercialized worldwide with a close and distinct relationship with its geographical place of origin. Some of the most famous wines, because of their high market value, have been a target of fraudulent admixtures, which have been reported through several media sources (newspaper, television, and internet). Considering that wines with commercial value are associated to a production region having distinctive autochthonous properties, this type of fraudulent practice is particularly important [32–34]. Nowadays, some renewed sophisticated consumers are interested in high quality wines, strongly linked to their region of origin [13]. This has led to a challenging topic regarding wine authenticity, which aims to obtain a provenience of origin signature for such wines. The establishment of wine production and geographic provenance limits, related to the wine terroir, is one of the most important issues in wine quality control. The use of geographical indications allows producers to obtain market recognition and often a premium price [35]. The development of sophisticated analytical techniques that are suitable for determining the geographical origin is highly desirable to guarantee the authenticity and geographical traceability of wines.

The presence and concentration of certain trace elements reflect the geochemistry and geomorphology of the different ecosystems. Recent studies have established that the content of selected volatiles (e.g., alcohols, esters, aldehydes, and ketones), elements (e.g., $^{87}Sr/^{86}Sr$, $^{13}C/^{12}C$, $^{18}O/^{16}O$, ME, $(D/H)_1$, $(D/H)_2$, $^{207}Pb/^{206}Pb$, $^2H/^1H$), and classical parameters (e.g., % ethanol, pH, total acidity, volatile acidity, malic acid, fructose, tartaric acid, lactic acid, succinate, citric acid, glycerol, 2,3-butandiol, dry matter, and relative density) in wines reflect the soil type, the environmental growing conditions, and the manufacturing processes, allowing wine regionality discrimination [36]. Factors, such as the amount of rainfall immediately prior to grape harvest (fermentation), and winery equipment, were shown to have a significant effect on the multi-element and multi-isotopic ratio and, consequently, were specific to the geographical origin of the wine [37].

Nowadays, the most established analytical methods are based on the profiling of trace elements (widely used for geographical discrimination), volatile compounds (used for varieties characterization), phenolic compounds (used for both varietal and geographical characterization, such as: Gallic, protocatechuic, vanillic, syringic, caffeic, p-coumaric and ferulic acids, catechin, epicatechin, quercetin, quercitrin, myricetin, kaempferol, and syringic and protocatechuic aldehydes) [38], organic constituents, mineral contents or composition, and light- or heavy-element isotope ratios using different chromatographic and spectroscopic methods [2]. In the last few years, there has been growing interest in developing analytical methods for wine-growing region authentication (Table 1).

Some of the most widely applied methods to assess the botanical and geographical origin of wine are spectroscopic and/or spectrometric, such as ultra-performance liquid chromatography (UPLC), Fourier transform ion cyclotron resonance mass spectrometry (FT-ICR-MS), nuclear magnetic resonance (NMR), ultraviolet-visible spectrophotometry (UV-vis), near-infrared (NIR), mid-infrared (MIR), high-performance liquid chromatography (HPLC), gas chromatography mass spectrometry (GC-MS), inductively coupled plasma optical emission spectrometry (ICP-OES), and inductively coupled plasma mass spectrometry (ICP-MS), among others. All these are high-throughput approaches requiring the use of somewhat complex statistical analyses and, most of the time, a big data set from the defined region considering several production years and varieties to develop a reliable database [31].

The relative abundance of stable isotope ratios of individual elements can act as fingerprints that enable the tracing of the origin of elements in a substance [39]. Because the stable isotope ratios within environmental substances have strong regional variations that are commonly controlled by the underlying geology, this means that these elements can be used as traceability indexes to determine their origins [37,39]. Consequently, stable isotopic ratio analysis of wine can allow its geographical origin to be authenticated thanks to the existence of an official European database (EU-Wine DB) [7]. Strontium isotopes reflect the local geological conditions of the wine terroir and may therefore be linked to the origin of the grapes used for wine production. The use of the ^{87}Sr/^{86}Sr isotope ratio as a geographical tracer of food origins is related to the constancy of its value in transferring from the soil to the plant and then into the final product. The Sr isotope ratios can be used to track the geographical origin of wines after analyzing soil, grape, and wine samples from producing areas [28,33]. A recent study involving several Cypriot wines was designed to monitor variations in isotopes and elements' content, aiming to relate them to the grapevine variety, environmental factors, and provenance [20]. The study was able to set a serious of elements (Na, Cu, B, Mn, K, Mg, P, (D/H)$_{\text{II}}$, R, and δ^{18}O) that could somehow access varietal identification, since they were also dependent on the geo-climatic conditions [20]. Several studies, using multi-isotopic analysis, have been conducted on the provenance of wines. Day et al. [40] combined (D/H)$_2$ data with multi-element data from 165 authentic grape samples for differentiation of the principal wine production zones in France for the 1990 vintage. These studies were quite promising, however, this method requires the establishment of quite big data-sets to be implemented and on validated independent models. Other authors reported a clear discrimination between wine production regions, reinforcing the importance of Sr isotopes' signature to characterize wine terroirs, and as a robust fingerprint to trace the geographic authenticity of wine [41,42]. Microbial terroir likely involves multiple interactions and have demonstrated that grape and wine microbiota exhibit regional patterns that correlate with wine chemical composition, suggesting that the grape microbiome may influence terroir in aspects, such as microbial distribution, strain diversity, and plant-microbial interactions [32].

Nevertheless, the combination of different methods able to analyze different types of wine compounds seems to be the most promising approach to establish a wine's geographical origin ([35]; Table 1).

Table 1. Overview of analytical techniques for tracing the geographic provenance of wines.

Samples	Analytical Technique	Data Analyzed/Analyte	Purpose of Analysis	References
Mass Spectrometry				
Grape, wine, and soil	IR-MS	$^{87}SR/^{86}SR$	Geographic origin of wine from Canada	[33]
Rocks, soils, and wine	IR-MS	$^{87}SR/^{86}SR$	Geologic and pedologic traceability of Italian wines	[41]
Red wines, musts grape juices, soils, and rocks	IR-MS	$^{87}SR/^{86}SR$	Fingerprinting wine geographic provenance.	[42]
Musts, soils, and grape components (skin, seeds, must, and stem)	TIMS and XRD spectra	$^{87}SR/^{86}SR$	Geographic traceability study of Italian white wine labelled with the Controlled Designation of Origin (DOC)	[36]
Sparkling wines	IR-MS	$\delta^{13}C$	The $\delta^{13}C$ evaluation in the sparkling wines to detect adulteration—wines chaptalization	[43]
Wines and rocks	TIMS	$^{87}SR/^{86}SR$	Radiogenic isotopic evaluation for tracing geographic provenance of wines	[44]
Soils, grapes, and wines	AAS, IR-MS, MC-ICP-MS	$\delta^{18}O$, $(D/H)_I$, $(D/H)_{II}$, $\delta^{13}C$, $\delta^{15}N$, and $^{87}Sr/^{86}Sr$	Development of a geographical traceability model	[45]
Vineyard soils	ICP-MS	$^{87}SR/^{86}SR$	Evaluation of $^{87}Sr/^{86}Sr$ ratio in vineyard soils from Portuguese Denominations of Origin and its potential for origin authentication	[28]
Wines	ICP–MS and multi-element analysis	Li, B, Mg, Al, Si, Cl, Sc, Mn, Ni, Ga, Se, Rb, Sr, Nb, Cs, Ba, La, W, Tl, and U	South African wines classification according to geographical origin	[46]
Red, white, and palhete amphora wines	ICP-MS	Mineral content	Elemental composition characterization of Alentejo wines to establish the geographic origin	[47]
Wine	ICP-MS, ICP-OES and IRMS	Elemental profile (Ca, Al, Mg, B, Fe, K, Rb, Mn, Na, P, Co, Ga, As, Sr), and Isotope ratio ($\delta^{18}O$)	Geographical origin of Chinese wines	[48]
Soils, grapes, and wines	ICP-MS	Cr, Co, Ni, Ga, Se, Y, Zr, Nb, Mo, Pd, In, La, Pr, Sm, Eu, Gd, Tm, Yb, Au, Tl, Th, U	Elemental patterns of wines, grapes, and vineyard soils from Chinese wine-producing regions and their origin association	[49]
Monovarietal wines	ICP-MS	Ba, As, Pb, Mo, and Co	Geographical origin differentiation of Argentinean white wines by their elemental profile	[50]
Spectroscopy				
Wines	SNIF-NMR	Isotopic and trace elements	Characterization of the geographic origin of Bordeaux wines	[51]
Wines	IRMS and SNIF-NMR	Isotopic ratios hydrogen ($^{2}H/^{1}H$), carbon ($^{13}C/^{12}C$), nitrogen ($^{15}N/^{14}N$), oxygen ($^{18}O/^{16}O$)	Regional origin discrimination of Slovenian Wines	[52]
Wines	NMR and MS	Cd, Cr, Cs, Er, Ga, Mn, and Sr	Wine adulterations	[53]
Wines	SNIF-NMR and IRMS in combination with chemometric	Multielement analysis	Geographical origin	[54]

Table 1. *Cont.*

Samples	Analytical Technique	Data Analyzed/Analyte	Purpose of Analysis	References
	Spectroscopy			
Wines authentication	^{1}H NMR, ICP-AES, HPLC	^{1}H and ^{13}C	Classification of wines from Slovenia and from Apulia	[55]
Red wines	MIR	Multielement analysis	Discrimination of wines based on their geographical origin and vintage year	[56]
Red wines	NIR combined with multivariate analysis (PCA, PLS-DA, LDA)	Chemometrics	Geographic classification of Spanish and Australian tempranillo wines	[57]
Sweet wines	F-AAS	Metallic content (Na, K, Ca, Mg, Fe, and Cu)	Classification and geographical differentiation of wines from Canary Islands (Spain)	[58]
	Separation			
Red wines	HPLC, UV, and fluorescence detection	Polyphenol content	Polyphenolic compounds quantification to typify wines according to their geographical origin	[38]
Red wines	HPLC-DAD	Polyphenolic components	Red wines differentiation based on cultivar and geographical origin with application of chemometrics of principal polyphenolic constituents	[59]
Monovarietal wines	HPLC	Non-flavonoid phenolic compounds: hydroxybenzoic acids, hydroxycinnamates, and Stilbenes	Czech Republic wines authentication: Wine discrimination according to the geographical origin	[60]
Red wines	RP-HPLC-DAD-F	Chromatographic profiles and chemometric data analysis	Classification and characterization of Spanish wines according to their appellation of origin.	[61]
Monovarietal red and white wines	SPME-MS and SPME-GC/MS	Volatiles compounds	Differentiation of wines according to grape variety and geographical origin	[62]
Red wines	HPLC	Organic acids (Shikimic and galacturonic acids); phenolic compounds (e.g., alkanes, aldehydes, alcohols, acids).	Varietal and geographic classification of wines according to their geographical origin	[63]
	HS-SPME GC×GC-TOFMS	Volatile compounds		[64]
Red wines	CE	Metals content (Na, K, Ca, Mg, Mn, and Li)	Wines classification according to their geographical origin.	[65]
	Others			
Must and grapes microbiota	DNA	High-throughput sequencing, molecular markers (SSR)	Biogeographical wines characteristics	[66]
Grapevines' fungal communities	DNA	Pyrosequencing of the 26S rRNA gene region	Vine fungi biogeography	[67]
Grape varieties	DNA	Ribosomal ITS region	Geographical region and grape varieties are drivers of population structures of fermentative vineyard-associated *S. cerevisiae* strains	[68]

Table 1. *Cont.*

Samples	Analytical Technique	Data Analyzed/Analyte	Purpose of Analysis	References
Others				
Grape yeast biota	DNA	RFLP and DNA sequencing	Azorean geographical indications wines: Grape-associated microbial biogeography from five islands of Azores Archipelago	[69]
Sensory				
Wine	Electronic nose and amperometric electronic tongue	Aroma	Characterization and classification of Italian Barbera wines	[70]
Wine	Electronic nose (fast gas chromatograph)	Aroma profile	Geographical classification of Chilean wines	[71]

IRMS—Isotope Ratio Mass Spectrometry; **ICP-MS**—Inductively coupled plasma mass spectrometry; **ICP-OES**—Inductively coupled plasma optical emission spectroscopy; **NMR**—Nuclear Magnetic Resonance spectroscopy; **SNIF-NMR**—Site-specific Natural Isotopic Fractionation; **FTIR**—Fourier transform Infrared; **MIR**—Mid-infrared spectroscopy; **NIR**—Near-infrared spectroscopy; **IR**—Infrared Spectroscopy; **HPLC-DAD**—High Performance Liquid Chromatography-Diode array detection; **GC**—Gas chromatography; **CE**—Capillary electrophoresis; **PCR-DNA**—Polymerase Chain Reaction based on DNA molecule; **RFLP**—Restriction Fragment Length Polymorphism Analysis; **ITS**—internal transcribed spacer; **SSR-SPME-GC-MS**—solid-phase microextraction-coupled to a gas chromatography-mass spectrometry; **GC-MS**—Gas chromatography mass spectrometry; **UV-VIS**—Ultraviolet and visible spectroscopy; **PCA**—Principal component analysis; **PLS-DA**—discriminant partial least-squares discriminant analysis; **LDA**—linear discriminant analysis; **F-AAS**—Flame-Atomic absorption spectroscopy; **TIMS**—thermal ionization mass spectrometry; **XRD**-ray powder diffraction.

4. DNA Fingerprinting for Varietal Identification

The assessment of a wine traceability and authenticity system embraces a huge and complex DNA-based techniques network (Figure 3) and requires a multidisciplinary analysis, including analytical and molecular validations. Unfortunately, the inconsistencies of the results obtained by analytical assays (e.g., protein, metabolite), due to environmental conditions and processing procedures, makes molecular DNA-based methods the preferred choice when dealing with grapevine varietal identification. In a food authentication molecular approach, there are several critical and important associated research areas, such as sampling and DNA extraction methodology, the development of specific molecular markers, and the sensitivity and suitableness of the detection method, that need to be considered.

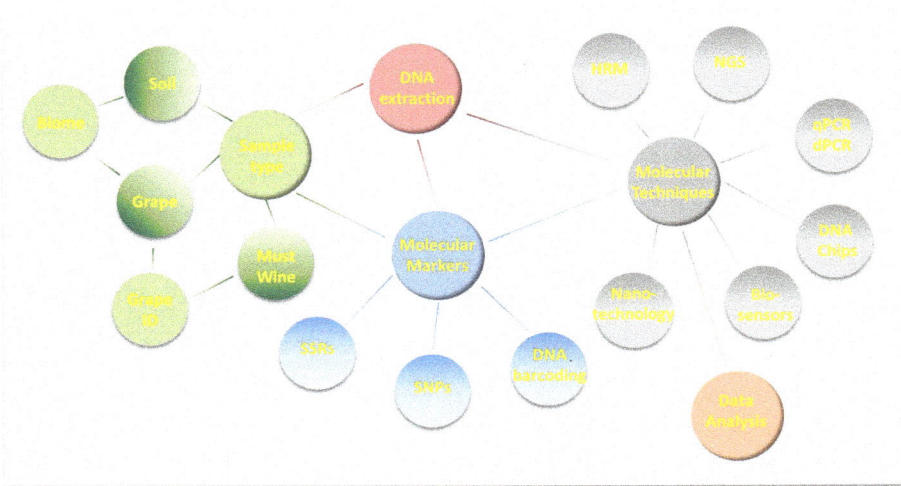

Figure 3. Wine authenticity networks. Diagram highlighting the critical and most important DNA based research areas in wine authentication HRM (high resolution melting), NGS—(next generation sequencing), qPCR/dPCR—quantitative/digital polymerase chain reaction, SSRs (simple sequence repeats), SNPs (single nucleotide polymorphism).

As previously mentioned, wine authenticity relies essentially on the determination of geographical origins and grapevine varietal composition by analytical and molecular methods, with the purpose of confirming the statements in the labels to assure the quality, typicality, and authenticity of the wine. It is well known that climate and biological factors (soil, grapevine variety, and fauna), as well as viticulture and enological procedures, are required to establish the concept of wine terroir. In this context, sampling plays a fundamental role in the wine authenticity network (Figure 3). The determination of the geographical origin is associated with the vineyard soil's isotopic profile by correlating trace elements values in wine and soil samples [28] and microbiome composition. Recent studies highlight the contribution of the autochthonous properties of vineyard microbiota in the winemaking process of a wine from a particular region [32,72].

The microflora of grapes is highly variable, mostly due to the influence of external factors, such as environmental parameters, geographical location, grapevine varieties, and the application of phytochemicals on the vineyards [73]. All these factors are responsible for the final characteristics of the wine, affecting the flavor and aroma attributes and, consequently, its final quality and value. In this case, sampling leaves or berries will provide identification of the grapevine varieties and the native microbial environment for wine authentication [74]. When must and wine samples are used in the authentication process, the recovery of grapevine DNA from such samples is a challenging procedure.

Nevertheless, due to recent technological advancements, it is possible to detect and discriminate the grapevine variety present in such sample types [4,8,75]. In an authenticity system, it is crucial to establish a framework concerning the most suitable sampling and storage conditions regarding the subsequent steps and the final defined purpose (Figure 3). Therefore, it is imperative that all the procedures are well established, producing reproducible results.

Soil, grape, and wine are complex samples that contain many interfering agents for molecular analysis, such as impurities, phenols, acids, metal ions, and salts. Therefore, choosing the best DNA extraction protocols is essential. There are several commercial kits that could be used for DNA extraction from soils, however, the extraction procedure must be adapted to the sample type to produce the best results [76]. DNA extraction from grapevines is well established from any part of the plant [77], and, currently, DNA extraction from the complex must/wine matrices has also been achieved [4,21,78,79]. The DNA extraction efficiency is highly dependent on the wine (e.g., type, age, alcohol percentage, winemaking process) used and therefore the DNA extraction protocol needs to be appropriately designed. In this process, wine properties must be considered, since the presence of natural compounds (e.g., phenolic, polysaccharides) and organic solvents (e.g., phenol/chloroform) used in the DNA extraction protocol may become problematic when proceeding with DNA analysis, as they can interfere with polymerase chain reaction (PCR) amplification. Thus, DNA extraction is one of the main and limiting steps, requiring the establishment of efficient DNA extraction protocols to ensure sufficient DNA yield, which will enable subsequent fingerprint analysis.

High throughput sequencing platforms have recently emerged and have been widely applied in the development of genomic markers. These methods vary in their applicability in terms of the research demands and molecular resources required. DNA markers offer an unequivocal and powerful tool towards a consistent wine authenticity system (Figure 3). The main drawback of DNA based technologies is the DNA degradation/fragmentation observed in processed food-samples and beverages, such as wine [21]. In this context, the quality and length of the DNA fragments retrieved are crucial to establish a proper fingerprint methodology. DNA recovered from wine samples is normally composed of very short length DNA molecules that result from DNA degradation caused by wine fermentation, aging, and storage. Usually, targeted fragments above 200–400 bp, using DNA extracted from must and wine samples as templates, are more difficult to amplify; however, they are still feasible in some regions [80]. When authentication is based on the use of SSR markers, a precise selection of SSR loci must be previously undertaken, considering the discriminative power of the marker (must be high), the molecular weight size range (must be low), the robustness, and the reproducibility. Furthermore, the establishment of a worldwide SSR database, allowing comparison of the results, will reduce the chance of ambiguous varietal identification.

Currently, advancements on sequencing technologies have allowed a wider SNP marker identification between and among different grapevine varieties [24]. SNPs' high abundance and wide distribution through the genome enable the amplification of very small fragments, thus being compatible with DNA recovered from must and wine samples. However, SNPs are, in most cases, biallelic, therefore, to have a reasonable discriminative power among the different profiles, a larger number of SNP markers are required in comparison to SSR markers. For grapevine discrimination, a set of 48 SNPs have been successfully applied to leaf samples [24].

DNA barcoding is useful in certifying both the origin and quality of raw materials, and to detect adulterations in the industrial food chain. In general, DNA barcoding is based on the amplification of short DNA fragments belonging to the mitochondrial (animal foodstuffs) or chloroplast (plant foodstuffs) genomes, which are conserved at the species levels and preserved in most of the processed food products, therefore, being advantageous when compared to other DNA fingerprinting and genotyping approaches. However, DNA barcoding has as its main limit low intraspecific polymorphism, compromising its capacity in distinguishing closely related species [81]. Therefore, the barcode has evolved through the reduction of long barcode regions to short subregions, allowing the species to still show enough divergence. DNA barcoding represents a well-proven molecular approach to assess the authenticity of food items, although its use is hampered by analytical constraints [82].

Nowadays, technologies based on genomics and bioinformatics approaches are considered the most efficient tools for assessing the genetic authenticity of food products, and, therefore, their incorporation in traceability systems is highly advantageous (Figure 3). Among the DNA-based technologies, High Resolution Melting (HRM) has been shown to be an interesting technology for food authenticity purposes [8]. The recent advances on the instrumentation utilized, as well as on specialized and more efficient fluorescent DNA-binding dyes, have allowed this technique to become a high-throughput screening assay for grapevine varietal identification and wine analysis [8,80,83].

Traditional methods are based on PCR amplification designed for a small number of targets. Usually, this type of approach requires prior knowledge of the target species. The results obtained by direct PCR detection produce presence/absence results for the targeted species, however, no additional information is obtained, such as the presence of other species in the sample. Next generation sequencing (NGS) appears to overcome this drawback. NGS analysis enables the identification of different species in complex food matrices based on the result of a single and unique DNA sequence. Currently, NGS is the only test method that ensures the correct identification of species in complex food matrices by comparison with databases (containing several thousands of species) [84]. Additionally, NGS techniques can also overcome the issue of DNA fragmentation caused during the food processing. An NGS approach can be optimized to target short fragments, thus avoiding false negative results. However, DNA sequences must be informative, resulting in a DNA barcode, since it is a unique identifier. For these reasons, the use of NGS technologies on degraded DNA for authentication purposes would be especially interesting in food analysis [85] and possibly in wine authenticity.

Real-time PCR is still the prime method for food analysis, including pathogen detection, allergens identification, and detection and quantification of different species. High sensitivity, specificity, and reproducibility, and low levels of cross-contamination and reduced analysis time makes real-time PCR an attractive and alternative method to conventional PCR [86]. However, the most important advantage of real-time PCR is its capacity to quantify the starting amount of a specific DNA target. Real-time PCR chemistries are classified into two main groups: Double stranded DNA intercalating molecules or binding dyes, such as SYBR green I and EvaGreen; and fluorophore-labeled oligonucleotides, such as TaqMan probes [86]. Real Time-PCR is being continuously improved on through its instrumentation and chemistry generating better signals, increased sensitivity, short detection times, and excellent stability without causing PCR inhibition. However, there are many challenges yet to be addressed. Recent progresses in RT-PCR analyses includes a new range of fluorescent probe chemistries and nanoparticles owing to their higher sensitivity and short detection times and microfluidic integrations, giving a promising outlook for gene-based point-of-care food analysis at a much lower cost [87]. In the wine sector, quantitative real-time PCR (qPCR) is being applied not only to identify and quantify total yeast population during fermentation and in wine samples to support the terroir concept [88], but also for genetic varietal discrimination and relative quantification in wine samples [22].

Future trends in food analysis will include digital PCR (dPCR). dPCR is an end-point technique that allows absolute quantification without the construction of standard curves. Briefly, the dPCR technique involves sub-dividing the DNA sample (with master-mix) into hundreds to thousands of individual units run concurrently with each other. The individual units are then treated either as negative reactions (no DNA target present) or positive reactions (DNA target presence) [87]. The fraction of negative reactions is used for absolute quantification of the initial DNA concentration of the sample as it follows the Poisson distributions. It is one of the most precise methods when dealing with DNA/RNA quantification. However, since it is a recent technique, further validation is required before it can become a viable replacement of RT-PCR as a standard method for the detection and quantification of DNA/RNA for food analysis.

DNA chips (DNA microarrays) may also be a valuable technique by proving to be a fast, reusable, continuous, selective, and sensitive detection system for fraudulent food products. DNA microarrays involve multiple species-specific oligonucleotide probes to produce distinct fluorescent patterns for the identification of different species providing a unique barcode fluorescent pattern for each species, enabling an effective food product authentication [89]. In wine research, DNA microarrays have been used is several studies, namely for the screening of wine yeast strains [90]. DNA chips are a promising technology that could enable the identification of yeast or bacteria strains linked to a specific region and winemaking practices. Furthermore, microarray technique has been applied to olive cultivars to assure olive oil authenticity and other food matrices [89,91].

DNA nanotechnology emerges as a powerful and growing research area in several fields, including food authenticity [92]. This new and promising technology functionally integrates DNA molecule and/or other nucleic acids with nanoparticles in different physicochemical forms to produce a range of composites with unique properties. These capabilities are attracting attention from food control research communities in pursuit of new applications, including (bio) sensing and labeling tools for the food sector, especially concerning safety and authenticity purposes [82,93]. The development of biosensors in response to this demand is seemingly promising [94]. Recent studies report the potential of a DNA-based biosensor for grapevine discrimination purposes [26,27,95]. This specific type of biosensor uses DNA strands as probes for sensing DNA targets and was developed based on the ability of single-strand DNA molecules to recognize and bind to their complementary strands in a sample. Using it as a transducer functionalized with the single-stranded DNA molecules, the biosensor can respond to alterations in the refractive index of the fiber's surrounding medium generated by analyte binding, and it will detect these interactions [96]. The biosensor proved to be able to distinguish specific grapevine varieties through the detection of small variations in a certain region of their genome using not only synthetic oligonucleotides, but also genomic DNA extracted from leaf, must, and wine samples [27]. The results are promising and show the potential of this technology to be applied to grapevine varietal fingerprinting throughout the wine-chain, analyzing DNA with different levels of contamination from matrices subjected to different processing levels, without the requirement of any labelling or PCR step [26,27].

The resulting data acquired from the above-mentioned DNA-based technologies require the application of various data analysis methods so the several datasets can be made understandable. The acquired data are complex and, therefore, to have a more comprehensive analysis, a multi-disciplinary approach using bioinformatics and data mining resources is required.

All scientific DNA methodologies/techniques presented herein offer a wide range of possibilities for the establishment of an accurate wine authentication system (Table 2). The analytical/molecular analysis, supported by scientific knowledge, current regulations, and by internationally documented quality standards, is required to protect consumers against fraudulent practices and ensure brand fair trade. Nevertheless, a continuous research effort is essential to address emerging wine origin/quality issues.

Table 2. Summary of the pros and cons of the DNA based techniques applied to wine authentication.

Method	Pros	Cons
HRM	• closed-tube method avoiding contaminations • high sensitivity • PCR products are analyzed without gels and hazardous chemicals • fast • data analysis can be performed automatically in a few minutes • allows a good species identification and differentiation • allow a high number of samples	• need for high-quality DNA extracts • dependent on the extraction method (presence of PCR inhibitors) • the results only produce presence/absence results for the targeted species • no quantification of nucleic acids occurs • no DNA quantification is performed • careful primer design required • specific software required
qPCR	• enables quantification of target DNA • measures PCR amplification (quantification of nucleic acids) as it occurs • no post-PCR processing	• is not used to identify the geographical origin of products, type of processing, or addition of chemical adulterant • Specific equipment required • Specific types of chemistries required (Taqman, SYBR green)
dPCR	• provides an absolute quantification of nucleic acids • more accurate and sensitive measurement of the number of copies of target DNA, especially for low concentration and mixed samples • ability to analyze samples containing species mixtures with high sensitivity and in a single trial, performing multiple reactions in parallel • can be performed in microarray format, which can potentially increase the sensitivity • efficient even if the copy number of the target is low and/or PCR inhibitors are present • no need to rely on references or standards	• low equipment offer • specific and expensive equipment required
NGS	• ensures the correct and unambiguous detection and identification of species • allows untargeted detection of thousands of organisms with no requirement for previous knowledge of the sample	• databases required • damage DNA requires a unique identifier (DNA barcode) • only provides relative information on the abundance of each species • specific equipment and software for data analysis required • technically challenging
Biosensors Nanotechnology DNA chips	• These items are being continuous developed for authenticity purposes. Future devices must link high performance (particularly high sensitivity and selectivity), higher number of samples, sequencing-free, faster detection, miniaturization, portability, and low cost. The design of such powerful devices requires innovative efforts, combining fundamental biological, chemical, and material sciences.	
DNA markers	• allow species identification and differentiation • stability under environmental conditions and production procedures • reliable and accurate for botanical and geographical origin	

35

Table 2. *Cont.*

Method	Pros	Cons
SSR	• high specificity allowing unequivocal species identification and differentiation • high reproducibility • highly informative	• labelled primers required • large consumable requirement • sequencer required • limited targets • databases required
SNP	• highly informative • high frequency of occurrence • highly reproducible • the analysis can be automated • allow species identification and differentiation according to the target DNA	• primer design required • relatively expensive • specific equipment required • databases required
DNA barcoding	• highly informative • allow species identification and differentiation • use short DNA sequences from the standard part of the genome for species identification overcoming DNA fragmentation	• careful primer design required • only provide insights into species-level • databases required
Data analysis	• integrate a huge amount of biological data through data mining approaches and exploit such information by identifying statistically informative annotations	• specialized laboratories and equipment's required • skilled personnel required • databases required

5. Conclusions and Future Trends

The general food industry is searching for alternative methods applied to food monitoring and authentication. The implemented wine traceability system is not capable of efficiently controlling the production chain, and therefore requires urgent measures to reassure producers, retailers, and consumers against fraudulent practice. The development of alternative technological solutions supporting this have emerged throughout the years, giving a new insight into the sector. However, none of the developed technologies can tackle the authentication of the wine terroir in all its dimensions. Nonetheless, a multidisciplinary approach can be developed, aiming to tackle the main features of the terroir (geographical origin and grapevine varietal origin). Some of the possible technological approaches have been presented and should be considered in the future so that a robust traceability system may be designed for the wine sector.

Author Contributions: Authors P.M.-L. Conceptualization; and L.P., S.G., S.B., E.P.G., M.B.-C., J.R.F., P.M.-L. Writing-Review & Editing the manuscript.

Funding: This research was funded by the Norte 2020 through the project "INNOVINE&WINE-NORTE-01-0145-FEDER-000038" and the Portuguese Foundation for Science and Technology in the project "WineBioCode PTDC/AGR-ALI/117341/2010-FCOMP-01-0124-FEDER-019439", and postdoctoral grants, S.G. [BPD/UTAD/INNOVINE&WINE/457/2016] and L.P. [SFRH/BPD/123934/2016].

Conflicts of Interest: The authors declare no conflict of interest.

References

1. Arvanitoyannis, I.S. Wine authenticity, traceability and safety monitoring. In *Managing Wine Quality*; Reynolds, A., Ed.; Woodhead: New York, NY, USA, 2010; pp. 218–270, ISBN 978-1-84569-484-5.
2. Versari, A.; Laurie, V.F.; Ricci, A.; Laghi, L.; Parpinello, G.P. Progress in authentication, typification and traceability of grapes and wines by chemometric approaches. *Food Res. Int.* **2014**, *60*, 2–18. [CrossRef]
3. Palade, M.; Popa, M.-E. Wine traceability and authenticity—A literature review. *Sci. Bull. Ser. F Biotechnol.* **2014**, *18*, 226–233.
4. Pereira, L.; Martins-Lopes, P.; Batista, C.; Zanol, G.C.; Clímaco, P.; Brazão, J.; Eiras-Dias, J.E.; Guedes-Pinto, H. Molecular markers for assessing must varietal origin. *Food Anal. Methods* **2012**, *5*, 1252–1259. [CrossRef]
5. Moreno-Arribas, M.V.; Polo, M.C. Winemaking biochemistry and microbiology: Current knowledge and future trends. *Crit. Rev. Food Sci. Nutr.* **2005**, *45*, 265–286. [CrossRef] [PubMed]
6. Vukatana, K.; Sevrani, K.; Hoxha, E. Wine traceability: A data model and prototype in albanian context. *Foods* **2016**, *5*, 11. [CrossRef] [PubMed]
7. Villano, C.; Lisanti, M.T.; Gambuti, A.; Vecchio, R.; Moio, L.; Frusciante, L.; Aversano, R.; Carputo, D. Wine varietal authentication based on phenolics, volatiles and DNA markers: State of the art, perspectives and drawbacks. *Food Control* **2017**, *80*, 1–10. [CrossRef]
8. Pereira, L.; Gomes, S.; Barrias, S.; Fernandes, J.R.; Martins-Lopes, P. Applying high-resolution melting (HRM) technology to olive oil and wine authenticity. *Food Res. Int.* **2018**, *103*, 170–181. [CrossRef] [PubMed]
9. Sotirchos, D.G.; Danezis, G.P.; Georgiou, C.A. Introduction, Definitions and Legislation. In *Food Authentication: Management, Analysis and Regulation*; Georgiou, C.A., Danezis, G.P., Eds.; Wiley-Blackwell: West Sussex, UK, 2017; pp. 3–18, ISBN 978-1-118-81026-2.
10. OIV. *International Standard for the Labelling of Wines*; Organisation Internationale de la Vigne et du Vin: Paris, France, 2015.
11. Vineyards. Available online: https://vineyards.com (accessed on 31 July 2018).
12. OIV. *Distribution of the World's Grapevine Varieties*; Organisation Internationale de la Vigne et du Vin: Paris, France, 2017.
13. Danezis, G.P.; Tsagkaris, A.S.; Camin, F.; Brusic, V.; Georgiou, C.A. Food authentication: Techniques, trends & emerging approaches. *Trends Anal. Chem.* **2016**, *85*, 123–132. [CrossRef]
14. Basalekou, M.; Stratidaki, A.; Pappas, C.; Tarantilis, P.; Kotseridis, Y.; Kallithraka, S. Authenticity determination of greek-cretan mono-varietal white and red wines based on their phenolic content using Attenuated Total Reflectance Fourier Transform Infrared spectroscopy and chemometrics. *Curr. Res. Nutr. Food Sci. J.* **2016**, *4*, 54–62. [CrossRef]

15. Lohumi, S.; Lee, S.; Lee, H.; Cho, B.-K. A review of vibrational spectroscopic techniques for the detection of food authenticity and adulteration. *Trends Food Sci. Technol.* **2015**, *46*, 85–98. [CrossRef]
16. Perestrelo, R.; Silva, C.; Câmara, J.S. A useful approach for the differentiation of wines according to geographical origin based on global volatile patterns. *J. Sep. Sci.* **2014**, *37*, 1974–1981. [CrossRef] [PubMed]
17. Duchowicz, P.R.; Giraudo, M.A.; Castro, E.A.; Pomilio, A.B. Amino acid profiles and quantitative structure-property relationship models as markers for Merlot and Torrontés wines. *Food Chem.* **2013**, *140*, 210–216. [CrossRef] [PubMed]
18. Rešetar, D.; Marchetti-Deschmann, M.; Allmaier, G.; Katalinić, J.P.; Kraljević Pavelić, S. Matrix assisted laser desorption ionization mass spectrometry linear time-of-flight method for white wine fingerprinting and classification. *Food Control* **2016**, *64*, 157–164. [CrossRef]
19. González-Neves, G.; Favre, G.; Piccardo, D.; Gil, G. Anthocyanin profile of young red wines of Tannat, Syrah and Merlot made using maceration enzymes and cold soak. *Int. J. Food Sci. Technol.* **2016**, *51*, 260–267. [CrossRef]
20. Kokkinofta, R.; Fotakis, C.; Zervou, M.; Zoumpoulakis, P.; Savvidou, C.; Poulli, K.; Louka, C.; Economidou, N.; Tzioni, E.; Damianou, K.; et al. Isotopic and elemental authenticity markers: A case study on cypriot wines. *Food Anal. Methods* **2017**, *10*, 3902–3913. [CrossRef]
21. Pereira, L.; Guedes-Pinto, H.; Martins-Lopes, P. An enhanced method for *Vitis vinifera* L. DNA extraction from wines. *Am. J. Enol. Vitic.* **2011**, *62*, 547–552. [CrossRef]
22. Catalano, V.; Moreno-Sanz, P.; Lorenzi, S.; Grando, M.S. Experimental review of DNA-based methods for wine traceability and development of a single-nucleotide polymorphism (SNP) genotyping assay for quantitative varietal authentication. *J. Agric. Food Chem.* **2016**, *64*, 6969–6984. [CrossRef] [PubMed]
23. OIV. *2nd Edition of the OIV Descriptor List for Grape Varieties and Vitis Species*; OIV: Paris, France, 2007.
24. Cabezas, J.A.; Ibáñez, J.; Lijavetzky, D.; Vélez, D.; Bravo, G.; Rodríguez, V.; Carreño, I.; Jermakow, A.M.; Carreño, J.; Ruiz-García, L.; et al. A 48 SNP set for grapevine cultivar identification. *BMC Plant Biol.* **2011**, *11*, 153. [CrossRef] [PubMed]
25. REGULATION (EC) No 178/2002 of the European Parliament and of the Council of 28 January 2002 laying down the general principles and requirements of food law, establishing the European Food Safety Authority and laying down procedures in matters of food safety. *Off. J. Eur. Commun.* **2002**, *31*, 1–24.
26. Gomes, S.; Castro, C.; Barrias, S.; Pereira, L.; Jorge, P.; Fernandes, J.R.; Martins-Lopes, P. Alternative SNP detection platforms, HRM and biosensors, for varietal identification in *Vitis vinifera* L. using F3H and LDOX genes. *Sci. Rep.* **2018**, *8*, 5850. [CrossRef] [PubMed]
27. Barrias, S.; Fernandes, J.R.; Eiras-Dias, J.E.; Brazão, J.; Martins-Lopes, P. Label free DNA-based optical biosensor as a potential system for wine authenticity. *Food Chem.* **2019**, *270*, 299–304. [CrossRef] [PubMed]
28. Martins, P.; Madeira, M.; Monteiro, F.; Bruno de Sousa, R.; Curvelo-Garcia, A.S.; Catarino, S. ^{87}Sr/^{86}Sr ratio in vineyard soils from Portuguese denominations of origin and its potential for origin authentication. *J. Int. Sci. Vigne Vin* **2014**, *48*, 21–29. [CrossRef]
29. Esslinger, S.; Riedl, J.; Fauhl-Hassek, C. Potential and limitations of non-targeted fingerprinting for authentication of food in official control. *Food Res. Int.* **2014**, *60*, 189–204. [CrossRef]
30. Fang, F.; Li, J.-M.; Zhang, P.; Tang, K.; Wang, W.; Pan, Q.-H.; Huang, W.-D. Effects of grape variety, harvest date, fermentation vessel and wine ageing on flavonoid concentration in red wines. *Food Res. Int.* **2008**, *41*, 53–60. [CrossRef]
31. Fernandes, J.R.; Pereira, L.; Jorge, P.; Moreira, L.; Gonçalves, H.; Coelho, L.; Alexandre, D.; Eiras-Dias, J.; Brazão, J.; Clímaco, P.; et al. Wine fingerprinting using a bio-geochemical approach. *BIO Web Conf.* **2015**, *5*, 02021. [CrossRef]
32. Bokulich, N.A.; Collins, T.S.; Masarweh, C.; Allen, G.; Heymann, H.; Ebeler, S.E.; Mills, D.A. Associations among wine grape microbiome, metabolome, and fermentation behavior suggest microbial contribution to regional wine characteristics. *MBio* **2016**, *7*, e00631-16. [CrossRef] [PubMed]
33. Vinciguerra, V.; Stevenson, R.; Pedneault, K.; Poirier, A.; Hélie, J.-F.; Widory, D. Strontium isotope characterization of wines from Quebec, Canada. *Food Chem.* **2016**, *210*, 121–128. [CrossRef] [PubMed]
34. Belda, I.; Zarraonaindia, I.; Perisin, M.; Palacios, A.; Acedo, A. From vineyard soil to wine fermentation: microbiome approximations to explain the "terroir" concept. *Front. Microbiol.* **2017**, *8*, 821. [CrossRef] [PubMed]

35. Luykx, D.M.A.M.; van Ruth, S.M. An overview of analytical methods for determining the geographical origin of food products. *Food Chem.* **2008**, *107*, 897–911. [CrossRef]

36. Petrini, R.; Sansone, L.; Slejko, F.F.; Buccianti, A.; Marcuzzo, P.; Tomasi, D. The ^{87}Sr/^{86}Sr strontium isotopic systematics applied to Glera vineyards: A tracer for the geographical origin of the Prosecco. *Food Chem.* **2015**, *170*, 138–144. [CrossRef] [PubMed]

37. Kelly, S.; Heaton, K.; Hoogewerff, J. Tracing the geographical origin of food: The application of multi-element and multi-isotope analysis. *Trends Food Sci. Technol.* **2005**, *16*, 555–567. [CrossRef]

38. Rodríguez-Delgado, M.-Á.; González-Hernández, G.; Conde-González, J.-E.; Pérez-Trujillo, J.-P. Principal component analysis of the polyphenol content in young red wines. *Food Chem.* **2002**, *78*, 523–532. [CrossRef]

39. Nakano, T. Potential uses of stable isotope ratios of Sr, Nd, and Pb in geological materials for environmental studies. *Proc. Jpn. Acad. Ser. B* **2016**, *92*, 167–184. [CrossRef] [PubMed]

40. Day, M.P.; Zhang, B.; Martin, G.J. Determination of the geographical origin of wine using joint analysis of elemental and isotopic composition. II—Differentiation of the principal production zones in france for the 1990 vintage. *J. Sci. Food Agric.* **1995**, *67*, 113–123. [CrossRef]

41. Braschi, E.; Marchionni, S.; Priori, S.; Casalini, M.; Tommasini, S.; Natarelli, L.; Buccianti, A.; Bucelli, P.; Costantini, E.A.C.; Conticelli, S. Tracing the ^{87}Sr/^{86}Sr from rocks and soils to vine and wine: An experimental study on geologic and pedologic characterisation of vineyards using radiogenic isotope of heavy elements. *Sci. Total Environ.* **2018**, *628–629*, 1317–1327. [CrossRef] [PubMed]

42. Marchionni, S.; Buccianti, A.; Bollati, A.; Braschi, E.; Cifelli, F.; Molin, P.; Parotto, M.; Mattei, M.; Tommasini, S.; Conticelli, S. Conservation of ^{87}Sr/^{86}Sr isotopic ratios during the winemaking processes of "Red" wines to validate their use as geographic tracer. *Food Chem.* **2016**, *190*, 777–785. [CrossRef] [PubMed]

43. Martinelli, L.A.; Moreira, M.Z.; Ometto, J.P.H.B.; Alcarde, A.R.; Rizzon, L.A.; Stange, E.; Ehleringer, J.R. Stable carbon isotopic composition of the wine and CO_2 bubbles of sparkling wines: Detecting C_4 sugar additions. *J. Agric. Food Chem.* **2003**, *51*, 2625–2631. [CrossRef] [PubMed]

44. Marchionni, S.; Braschi, E.; Tommasini, S.; Bollati, A.; Cifelli, F.; Mulinacci, N.; Mattei, M.; Conticelli, S. High-precision ^{87}Sr/^{86}Sr analyses in wines and their use as a geological fingerprint for tracing geographic provenance. *J. Agric. Food Chem.* **2013**, *61*, 6822–6831. [CrossRef] [PubMed]

45. Durante, C.; Baschieri, C.; Bertacchini, L.; Bertelli, D.; Cocchi, M.; Marchetti, A.; Manzini, D.; Papotti, G.; Sighinolfi, S. An analytical approach to Sr isotope ratio determination in Lambrusco wines for geographical traceability purposes. *Food Chem.* **2015**, *173*, 557–563. [CrossRef] [PubMed]

46. Coetzee, P.P.; Steffens, F.E.; Eiselen, R.J.; Augustyn, O.P.; Balcaen, L.; Vanhaecke, F. Multi-element analysis of South African wines by ICP–MS and their classification according to geographical origin. *J. Agric. Food Chem.* **2005**, *53*, 5060–5066. [CrossRef] [PubMed]

47. Cabrita, M.J.; Martins, N.; Barrulas, P.; Garcia, R.; Dias, C.B.; Pérez-Álvarez, E.P.; Costa Freitas, A.M.; Garde-Cerdán, T. Multi-element composition of red, white and palhete amphora wines from Alentejo by ICPMS. *Food Control* **2018**, *92*, 80–85. [CrossRef]

48. Fan, S.; Zhong, Q.; Gao, H.; Wang, D.; Li, G.; Huang, Z. Elemental profile and oxygen isotope ratio (δ^{18}O) for verifying the geographical origin of Chinese wines. *J. Food Drug Anal.* **2018**, *26*, 1033–1044. [CrossRef] [PubMed]

49. Zou, J.-F.; Peng, Z.-X.; Du, H.-J.; Duan, C.-Q.; Reeves, M.J.; Pan, Q.-H. Elemental patterns of wines, grapes, and vineyard soils from Chinese wine-producing regions and their association. *Am. J. Enol. Vitic.* **2012**, *63*, 232–240. [CrossRef]

50. Azcarate, S.M.; Martinez, L.D.; Savio, M.; Camiña, J.M.; Gil, R.A. Classification of monovarietal Argentinean white wines by their elemental profile. *Food Control* **2015**, *57*, 268–274. [CrossRef]

51. Martin, G.J.; Mazure, M.; Jouitteau, C.; Martin, Y.-L.; Aguile, L.; Allain, P. Characterization of the geographic origin of Bordeaux wines by a combined use of isotopic and trace element Measurements. *Am. J. Enol. Vitic.* **1999**, *50*, 409–417.

52. Ogrinc, N.; Košir, I.J.; Kocjančič, M.; Kidrič, J. Determination of authenticy, regional origin, and vintage of Slovenian wines using a combination of IRMS and SNIF-NMR analyses. *J. Agric. Food Chem.* **2001**, *49*, 1432–1440. [CrossRef] [PubMed]

53. Ogrinc, N.; Košir, I.J.; Spangenberg, J.E.; Kidrič, J. The application of NMR and MS methods for detection of adulteration of wine, fruit juices, and olive oil. A review. *Anal. Bioanal. Chem.* **2003**, *376*, 424–430. [CrossRef] [PubMed]

54. Košir, I.J.; Kocjančič, M.; Ogrinc, N.; Kidrič, J. Use of SNIF-NMR and IRMS in combination with chemometric methods for the determination of chaptalisation and geographical origin of wines (the example of Slovenian wines). *Anal. Chim. Acta* **2001**, *429*, 195–206. [CrossRef]

55. Brescia, M.A.; Košir, I.J.; Caldarola, V.; Kidrič, J.; Sacco, A. Chemometric classification of apulian and slovenian wines using ^{1}H NMR and ICP-OES together with HPICE data. *J. Agric. Food Chem.* **2003**, *51*, 21–26. [CrossRef] [PubMed]

56. Picque, D.; Cattenoz, T.; Corrieu, G.; Berger, J. Discrimination of red wines according to their geographical origin and vintage year by the use of mid-infrared spectroscopy. *Sci. Aliments* **2005**, *25*, 207–220. [CrossRef]

57. Liu, L.; Cozzolino, D.; Cynkar, W.U.; Gishen, M.; Colby, C.B. Geographic classification of spanish and australian Tempranillo red wines by visible and near-infrared spectroscopy combined with multivariate analysis. *J. Agric. Food Chem.* **2006**, *54*, 6754–6759. [CrossRef] [PubMed]

58. Frías, S.; Trujillo, J.P.; Peña, E.; Conde, J.E. Classification and differentiation of bottled sweet wines of Canary Islands (Spain) by their metallic content. *Eur. Food Res. Technol.* **2001**, *213*, 145–149. [CrossRef]

59. Makris, D.P.; Kallithraka, S.; Mamalos, A. Differentiation of young red wines based on cultivar and geographical origin with application of chemometrics of principal polyphenolic constituents. *Talanta* **2006**, *70*, 1143–1152. [CrossRef] [PubMed]

60. Pavloušek, P.; Kumšta, M. Authentication of Riesling wines from the Czech Republic on the basis of the non-flavonoid phenolic compounds. *Czech J. Food Sci.* **2013**, *31*, 474–482. [CrossRef]

61. Serrano-Lourido, D.; Saurina, J.; Hernández-Cassou, S.; Checa, A. Classification and characterisation of Spanish red wines according to their appellation of origin based on chromatographic profiles and chemometric data analysis. *Food Chem.* **2012**, *135*, 1425–1431. [CrossRef] [PubMed]

62. Ziółkowska, A.; Wąsowicz, E.; Jeleń, H.H. Differentiation of wines according to grape variety and geographical origin based on volatiles profiling using SPME-MS and SPME-GC/MS methods. *Food Chem.* **2016**, *213*, 714–720. [CrossRef] [PubMed]

63. Etièvant, P.; Schlich, P.; Cantagrel, R.; Bertrand, M.; Bouvier, J.-C. Varietal and geographic classification of french red wines in terms of major acids. *J. Sci. Food Agric.* **1989**, *46*, 421–438. [CrossRef]

64. Robinson, A.L.; Adams, D.O.; Boss, P.K.; Heymann, H.; Solomon, P.S.; Trengove, R.D. Influence of geographic origin on the sensory characteristics and wine composition of *Vitis vinifera* cv. Cabernet Sauvignon wines from Australia. *Am. J. Enol. Vitic.* **2012**, *63*, 467–476. [CrossRef]

65. Peng, Y.; Liu, F.; Ye, J. Quantitative and qualitative analysis of flavonoid markers in Frucus aurantii of different geographical origin by capillary electrophoresis with electrochemical detection. *J. Chromatogr. B* **2006**, *830*, 224–230. [CrossRef] [PubMed]

66. Bokulich, N.A.; Ohta, M.; Richardson, P.M.; Mills, D.A. Monitoring seasonal changes in winery-resident microbiota. *PLoS ONE* **2013**, *8*, e66437. [CrossRef] [PubMed]

67. Taylor, M.W.; Tsai, P.; Anfang, N.; Ross, H.A.; Goddard, M.R. Pyrosequencing reveals regional differences in fruit-associated fungal communities. *Environ. Microbiol.* **2014**, *16*, 2848–2858. [CrossRef] [PubMed]

68. Schuller, D.; Cardoso, F.; Sousa, S.; Gomes, P.; Gomes, A.C.; Santos, M.A.S.; Casal, M. Genetic diversity and population structure of Saccharomyces cerevisiae strains isolated from different grape varieties and winemaking regions. *PLoS ONE* **2012**, *7*, e32507. [CrossRef] [PubMed]

69. Drumonde-Neves, J.; Franco-Duarte, R.; Lima, T.; Schuller, D.; Pais, C. Association between grape yeast communities and the vineyard ecosystems. *PLoS ONE* **2017**, *12*, e0169883. [CrossRef] [PubMed]

70. Buratti, S.; Benedetti, S.; Scampicchio, M.; Pangerod, E.C. Characterization and classification of Italian Barbera wines by using an electronic nose and an amperometric electronic tongue. *Anal. Chim. Acta* **2004**, *525*, 133–139. [CrossRef]

71. Beltrán, N.H.; Duarte-Mermoud, M.A.; Muñoz, R.E. Geographical classification of Chilean wines by an electronic nose. *Int. J. Wine Res.* **2009**, *2009*, 209–219. [CrossRef]

72. Knight, S.; Klaere, S.; Fedrizzi, B.; Goddard, M.R. Regional microbial signatures positively correlate with differential wine phenotypes: Evidence for a microbial aspect to terroir. *Sci. Rep.* **2015**, *5*, 14233. [CrossRef] [PubMed]

73. Pinto, C.; Pinho, D.; Cardoso, R.; Custódio, V.; Fernandes, J.; Sousa, S.; Pinheiro, M.; Egas, C.; Gomes, A.C. Wine fermentation microbiome: A landscape from different Portuguese wine appellations. *Front. Microbiol.* **2015**, *6*, 905. [CrossRef] [PubMed]

74. Mezzasalma, V.; Ganopoulos, I.; Galimberti, A.; Cornara, L.; Ferri, E.; Labra, M. Poisonous or non-poisonous plants? DNA-based tools and applications for accurate identification. *Int. J. Leg. Med.* **2017**, *131*, 1–19. [CrossRef] [PubMed]

75. Bigliazzi, J.; Scali, M.; Paolucci, E.; Cresti, M.; Vignani, R. DNA extracted with optimized protocols can be genotyped to reconstruct the varietal composition of monovarietal wines. *Am. J. Enol. Vitic.* **2012**, *63*, 568–573. [CrossRef]

76. Fatima, F.; Pathak, N.; Rastogi Verma, S. An improved method for soil DNA extraction to study the microbial assortment within rhizospheric region. *Mol. Biol. Int.* **2014**, *2014*. [CrossRef] [PubMed]

77. Lodhi, M.A.; Ye, G.-N.; Weeden, N.F.; Reisch, B.I. A simple and efficient method for DNA extraction from grapevine cultivars and *Vitis* species. *Plant Mol. Biol. Rep.* **1994**, *12*, 6–13. [CrossRef]

78. Baleiras-Couto, M.M.; Eiras-Dias, J.E. Detection and identification of grape varieties in must and wine using nuclear and chloroplast microsatellite markers. *Anal. Chim. Acta* **2006**, *563*, 283–291. [CrossRef]

79. Işçi, B.; Kalkan Yildirim, H.; Altindisli, A. Evaluation of methods for DNA extraction from must and wine. *J. Inst. Brew.* **2014**, *120*, 238–243. [CrossRef]

80. Pereira, L.; Gomes, S.; Castro, C.; Eiras-Dias, J.E.; Brazão, J.; Graça, A.; Fernandes, J.R.; Martins-Lopes, P. High Resolution Melting (HRM) applied to wine authenticity. *Food Chem.* **2017**, *216*, 80–86. [CrossRef] [PubMed]

81. Barcaccia, G.; Lucchin, M.; Cassandro, M. DNA barcoding as a molecular tool to track down mislabeling and food piracy. *Diversity* **2015**, *8*, 2. [CrossRef]

82. Valentini, P.; Galimberti, A.; Mezzasalma, V.; De Mattia, F.; Casiraghi, M.; Labra, M.; Pompa, P.P. DNA barcoding meets nanotechnology: Development of a universal colorimetric test for food authentication. *Angew. Chem. Int. Ed.* **2017**, *56*, 8094–8098. [CrossRef] [PubMed]

83. Pereira, L.; Martins-Lopes, P. *Vitis vinifera* L. Single-nucleotide polymorphism detection with high-resolution melting analysis based on the UDP-glucose:Flavonoid 3-O-Glucosyltransferase gene. *J. Agric. Food Chem.* **2015**, *63*, 9165–9174. [CrossRef] [PubMed]

84. Giusti, A.; Armani, A.; Sotelo, C.G. Advances in the analysis of complex food matrices: Species identification in surimi-based products using next generation sequencing technologies. *PLoS ONE* **2017**, *12*, e0185586. [CrossRef] [PubMed]

85. Tillmar, A.O.; Dell'Amico, B.; Welander, J.; Holmlund, G. A universal method for species identification of mammals utilizing next generation sequencing for the analysis of DNA mixtures. *PLoS ONE* **2013**, *8*, e83761. [CrossRef] [PubMed]

86. Navarro, E.; Serrano-Heras, G.; Castaño, M.J.; Solera, J. Real-time PCR detection chemistry. *Clin. Chim. Acta* **2015**, *439*, 231–250. [CrossRef] [PubMed]

87. Salihah, N.T.; Hossain, M.M.; Lubis, H.; Ahmed, M.U. Trends and advances in food analysis by real-time polymerase chain reaction. *J. Food Sci. Technol.* **2016**, *53*, 2196–2209. [CrossRef] [PubMed]

88. Wang, C.; García-Fernández, D.; Mas, A.; Esteve-Zarzoso, B. Fungal diversity in grape must and wine fermentation assessed by massive sequencing, quantitative PCR and DGGE. *Front. Microbiol.* **2015**, *6*, 1156. [CrossRef] [PubMed]

89. Scarano, D.; Rao, R. DNA markers for food products authentication. *Diversity* **2014**, *6*, 579–596. [CrossRef]

90. Mendes-Ferreira, A.; lí del Olmo, M.; García-Martínez, J.; Pérez-Ortín, J.E. Functional genomics in wine yeast: DNA arrays and next generation sequencing. In *Biology of Microorganisms on Grapes, in Must and in Wine*; König, H., Unden, G., Fröhlich, J., Eds.; Springer International Publishing: Cham, Switzerland, 2017; pp. 573–604, ISBN 978-3-319-60021-5. [CrossRef]

91. Consolandi, C.; Palmieri, L.; Severgnini, M.; Maestri, E.; Marmiroli, N.; Agrimonti, C.; Baldoni, L.; Donini, P.; De Bellis, G.; Castiglioni, B. A procedure for olive oil traceability and authenticity: DNA extraction, multiplex PCR and LDR–universal array analysis. *Eur. Food Res. Technol.* **2008**, *227*, 1429–1438. [CrossRef]

92. Nummelin, S.; Kommeri, J.; Kostiainen, M.A.; Linko, V. Evolution of structural DNA nanotechnology. *Adv. Mater.* **2018**, *30*, 1703721. [CrossRef] [PubMed]

93. Stephen Inbaraj, B.; Chen, B.H. Nanomaterial-based sensors for detection of foodborne bacterial pathogens and toxins as well as pork adulteration in meat products. *J. Food Drug Anal.* **2016**, *24*, 15–28. [CrossRef] [PubMed]

94. Mehrotra, P. Biosensors and their applications—A review. *J. Oral Biol. Craniofacial Res.* **2016**, *6*, 153–159. [CrossRef] [PubMed]

Beverages **2018**, *4*, 71

95. Gonçalves, H.M.R.; Moreira, L.; Pereira, L.; Jorge, P.; Gouveia, C.; Martins-Lopes, P.; Fernandes, J.R.A. Biosensor for label-free DNA quantification based on functionalized LPGs. *Biosens. Bioelectron.* **2016**, *84*, 30–36. [CrossRef] [PubMed]
96. Moreira, L.; Gonçalves, H.M.R.; Pereira, L.; Castro, C.; Jorge, P.; Gouveia, C.; Fernandes, J.R.; Martins-Lopes, P. Label-free optical biosensor for direct complex DNA detection using *Vitis vinifera* L. *Sens. Actuators B Chem.* **2016**, *234*, 92–97. [CrossRef]

beverages

MDPI

Article

Geographical Classification of Tannat Wines Based on Support Vector Machines and Feature Selection

Nattane Luíza Costa [1], Laura Andrea García Llobodanin [2], Inar Alves Castro [2] and Rommel Barbosa [1,*]

1   Institute of Informatics, Federal University of Goiás, 74690-900 Goiânia-Go, Brazil; nattaneluiza@hotmail.com
2   LADAF—Laboratory of Functional Foods, Department of Food and Experimental Nutrition, Faculty of Pharmaceutical Sciences, University of São Paulo, Av. Lineu Prestes 580, B14, 05508-900 São Paulo, Brazil; laugar18@hotmail.com (L.A.G.L.); inar@usp.br (I.A.C.)
*   Correspondence: rommel@inf.ufg.br; Tel.: +55-(62)-3521-1181; Fax: +55-(62)-3521-1182

Received: 30 September 2018; Accepted: 27 November 2018; Published: 30 November 2018

Abstract: Geographical product recognition has become an issue for researchers and food industries. One way to obtain useful information about the fingerprint of wines is by examining that fingerprint's chemical components. In this paper, we present a data mining and predictive analysis to classify Brazilian and Uruguayan Tannat wines from the South region using the support vector machine (SVM) classification algorithm with the radial basis kernel function and the *F*-score feature selection method. A total of 37 Tannat wines differing in geographical origin (9 Brazilian samples and 28 Uruguayan samples) were analyzed. We concluded that given the use of at least one anthocyanin (peon-3-glu) and the radical scavenging activity (DPPH), the Tannat wines can be classified with 94.64% accuracy and 0.90 Matthew's correlation coefficient (MCC). Furthermore, the combination of SVM and feature selection proved useful for determining the main chemical parameters that discriminate with regard to the origin of Tannat wines and classifying them with a high degree of accuracy. Additionally, to our knowledge, this is the first study to classify the Tannat wine variety in the context of two countries in South America.

Keywords: support vector machines; data mining; wine classification; Tannat wines; feature selection

---

1. Introduction

The contemporary global wine industry is inherently geographical, with the origins of the grapes being a main factor in the eminence of this particular beverage [1]. Additionally, customers are interested in finding high-quality wines that specify their geographical region [2]. In this regard, performing a fingerprinting approach to characterize wines and to classify their characteristics according to their production regions is a necessity. Tests to confirm the origin of a specific wine are necessary, because wine production and trade have always been associated with high costs [3]. Based on this background, detailed and continuous controls are essential for maintaining the quality of wine and for identifying possible cases of fraud with respect to their geographical origin [4].

Latin-American viticulture belong to the New World of wine, both in terms of productivity and levels of consumption [5]. Since the middle of the 19th century, Tannat has been the main variety of red wines cultivated in Uruguay [6]. This variety is adaptable to the ecological conditions of Uruguay, producing exceptional wines, highlighted by its originality [7]. In Brazil, a country new to wine production, Tannat is one of the six most cultivated varieties (along with Cabernet Franc, Cabernet Sauvignon, Merlot, Pinotage, Pinot Noir). Wine production has a significant economic and social impact on the southern region of Brazil, which represents 90% of wine production in that nation [8].

Unsupervised techniques, such as principal component analysis and cluster analysis, have commonly been employed to differentiate wines [9–13]. However, unsupervised pattern recognition methods must not be confused with classification methods. In some cases, these techniques allow for the identification and visualization of certain patterns, while in other cases, they may be insufficient for finding patterns between food profiles and physicochemical attributes, due to the multitude of compounds and physical attributes that are present in food products [14,15].

Other studies have used unsupervised techniques prior to using supervised techniques, such as linear discriminant analysis, partial least squares-discriminant analysis, and decision trees to classify wine samples [8,16–18]. In addition, other advanced techniques available through current data mining methodology have been successfully used to classify wines and other food products according to their geographical origin and variety, such as support vector machines (SVM) [19]. In addition, SVM is a suitable classifier for balanced datasets [20].

The SVM classification algorithm has been demonstrated to be useful for wine classification. For example, a total of 44 Polish white and red wine samples from different parts of Poland were classified [21]. The authors used principal component analysis (PCA) as an exploratory analysis and support vector machine method, with a radial basis function (RBF), to classify the data according to grape variety, geographic region, sugar content, alcohol content, type of yeast used, post-fermentation treatment, and the temperature in the fermentation process. The main results of this study of the classification of white and red wines, according to the variety of grapes used for production, showed 98.7 and 98.2% accuracy, respectively. SVM with an RBF kernel was compared to the partial least squares discrimination analysis (PLS-DA) to classify 1188 wine samples from South Africa, Hungary, Romania, and the Czech Republic [22]. Results showed that the SVM models, using 25 variables, have proven to be more efficient than the PLS-DA, with an accuracy of at least 95%. Another study compared SVM and PLS-DA in order to classify seventy-nine wine vinegar samples from three Spanish regions. The samples were classified according to their origin and their categories (aged and sweet), and the SVM classification models demonstrated a higher ability of prediction (between 92% and 100% correctly classified samples) than for the PLS-DA models [23].

A group of sixty-four samples of white wine belonging to four different Spanish DO (wine with designation of origin) were classified, based on SVM and linear discriminant analysis (LDA) [24]. The authors achieved an accuracy of 100% with the SVM model when using five selected variables, according to the Kruskal–Wallis test, the PCA, and the backward stepwise LDA. A previous study carried out by our group classified the geographical origin of Cabernet Sauvignon wines from Brazil and Chile with the use of SVM with an RBF kernel and correlation-based feature selection [25]. The results showed a good classification rate of 89% when using the 20 original elements, and an accuracy of 83% when using only five elements (L *, DPPH (2,2-diphenyl-1-picrylhydrazyl), delph-3-acetylglu, peon-3-(coum)glu, and pet-3-acetylglu). Another recent paper classified South America wines using SVM and other classifiers [26]. This study classified wines from Argentina, Chile, Brazil, and Uruguay based on their mineral content. The authors proposed a new feature selection method and had used SVM, LDA, neural networks, and Naïve Bayses, outperforming the other classifiers in all features, including the best feature, subset SVM.

The aim of this paper is to classify Tannat wines from the southern regions of Brazil and Uruguay, using data mining techniques. The contributions of this paper can be summarized as follows: 1) Analysis of the same wine variety from two different countries; 2) the possibility of irrelevant and redundant features of the wines' chemical parameters, as associated with their functionality, designed as antioxidant activity (DPPH and ORAC), total polyphenols (TP), total anthocyanins (TA) and color, is taken into consideration to select an optimal subset of features for classification; and 3) the use of data mining techniques to classify wines.

2. Materials and Methods

2.1. Wine Samples

Tannat wine samples were obtained from local markets and wine distributers in the city of São Paulo (Brazil). All the wines are monovarietal (at least 75% of Tannat variety), from 2009 and 2010 vintages, bottled in 750 mL bottles, and with retail prices between 1–50 United States dollars (USD). Samples were distributed as follows: Uruguay (*n* = 28) and Brazil (*n* = 9).

2.2. Chemical Compounds Determination

The variables analyzed for each wine were: Color by CIELAB (L *, a *, b *), total polyphenols (TPI), total anthocyanins (TA), antioxidant activity by oxygen radical absorbance capacity (ORAC) and free radical scavenging activity (DPPH), and individual anthocyanins (cyan-3-glu, delph-3-acetylglu, delph-3-glu, malv-3-(coum)glu, malv-3-acetylglu, malv-3-glu, peon-3-(coum)glu, peon-3-acetylglu, peon-3-glu, pet-3-(coum)glu, pet-3-acetylglu, pet-3-glu, and vitisin A).

2.3. Color Determination

The analysis of color was performed by measuring the transmittance in a ColorQuest XE colorimeter (Hunter Associates Laboratory, Inc., Reston, VA, USA) using the CIE 1964 standard observer (10° visual field) and the CIE standard illuminant D65 as references. The software EasyMatch QC (Hunter Associates Laboratory, Inc., Reston, VA, USA) was used to determine the three CIELAB coordinates: a * (red-green; +a *, −a *), b * (yellow-blue; +b *, −b *), and lightness L * (white-black, 0–100). The analyses were performed in triplicate.

2.4. Total Polyphenols

Total polyphenols (TPI) were determined using the Folin–Ciocalteu colorimetric method [27]. This method was adapted for measurement with a microplate reader, employing a standard curve of gallic acid (ranging from 0 to 100 mg/L). The results are expressed as mg gallic acid equivalents per liter (mg GAE/L). The analyses were performed in triplicate.

2.5. Total Monomeric Anthocyanins

Total anthocyanins (TA) were quantified, as described by Fuleki and Francis [28] with some modifications, using Equation (1):

$$TA = \frac{[(A510 - A700)pH\,1.0 - (A510 - A700)pH\,4.5] \times MV \times DF \times 103}{\varepsilon} \tag{1}$$

in which A510 and A700 are the absorbance at 510 and 700 nm, respectively, MW is the molecular weight of cyanidin-3-glucoside (449.2), DF is the dilution factor (10), and ε represents the molar absorptive of cyanidin-3-glucoside (26,900). The results are expressed as mg of cyanidin-3-glucoside per liter (mg cyanidin-3-glu/L). The experiments were performed in duplicate.

2.6. Individual Anthocyanins

2.6.1. HPLC–DAD

Individual anthocyanins were determined by HPLC, using the method described by Boido et al. [29] with some modifications. The identification of the anthocyanins was performed by comparing the retention time for each peak with available standards and values obtained from the literature. Analyses by HPLC–MS were performed to confirm these results, as described below. The individual anthocyanins were quantified using calibration curves of malvidin-3-glucoside (0.1–7.0 mg/L and 7.1–200 mg/L) and the areas obtained in the high-performance liquid chromatography with a diode-array detector

(HPLC-DAD) analysis. The results are expressed as milligrams of malvidin-3-glucoside per liter (mg malvidin-3-glu/L). The analyses were performed in duplicate.

2.6.2. HPLC–DAD–MS

A Shimadzu Prominence SPD-M20A liquid chromatograph (Shimadzu Co., Kyoto, Japan) that was connected via a UV cell outlet to a Bruker Esquire HCT ion trap mass spectrometer (BrukerDaltonics Inc., Billerica, MA, USA) was used for the analysis. The HPLC–DAD conditions were identical to those described in the previous section. The MS contained an electrospray ionization interface (ESI) and used nitrogen as the drying gas at a flow rate of 6.0 L/min. The pressure of the nebulizer was set at 25.0 psi. The capillary temperature was 280 °C. Spectra were recorded in a positive ion mode between m/z 50 and 1100. The mass spectrometer was programmed to perform a series of three consecutive scans: A full mass scan, an MS2 scan of the most abundant ions in the full mass scan, and an MS3 scan of the most abundant ions in the MS2 scan. The obtained data were analyzed using the Bruker Compass DataAnalysis 4.0 software (BrukerDaltonik GmbH, Bremen, Germany).

2.7. Antioxidant Activity

The in vitro antioxidant activity of the wines was determined using two methods: Measuring the free radical scavenging capacity (DPPH) and the oxygen radical absorbance capacity (ORAC).

2.7.1. Free Radical Scavenging Capacity (DPPH)

The method described by Arnous et al. [30] was used with some modifications. The antiradical activity (AAR) of each sample, expressed in millimolarTrolox equivalents (mM TRE), was calculated using Equation (2):

$$AAR = 0.872 \times In(\%\Delta A515) - 2.922, \tag{2}$$

where $In(\%\Delta A515) = \frac{[(A515,control - A515,t=150)]}{A515,control} \times 100$.

The absorbance was read at 515 nm using a UVmini-1240 UV-VIS spectrophotometer (Shimadzu Corporation, Kyoto, Japan). The equation was determined by linear regression ($r^2 = 0.992$) after the plotting of known Trolox solutions (0.25–1.50 mM) against concentration. The analyses were performed in duplicate.

2.7.2. Oxygen Radical Absorbance Capacity (ORAC)

The ORAC assay was performed according to the method described by Huang et al. [31] with slight modifications. A calibration curve was prepared using known Trolox solutions (6.25–100 μM). The results are expressed as micromolar Trolox equivalents (μM TRE). Each sample was analyzed in quadruplicate.

2.8. Data Mining

Data mining is the process of automatic discovery of useful information in a database using a series of steps that can be carried out to discern these patterns [32]. These steps can be summarized as data cleaning (to remove noise and inconsistent data); data selection (whereby data relevant to the analysis are selected from the database); data transformation (whereby data are transformed by performing summary or aggregation operations; feature selection; or feature extraction); data mining (whereby machine learning algorithms are performed); pattern evaluation (identification of the interesting patterns based on performance measures); and finally, knowledge presentation.

In this study, we performed the data mining study as presented in Figure 1. The dataset is composed of 21 columns and 37 rows, including 20 columns that represent the chemical compounds, one column representing the class label (Brazil or Uruguay) and each row representing a wine sample. We used the synthetic minority over sampling technique (SMOTE) to create 19 new samples for the Brazilian class because an imbalanced dataset can benefit the majority class [33]. SMOTE is the most widely and effectively used oversampling method; it creates synthetic minority class samples by interpolating between real minority examples and their nearest neighbors.

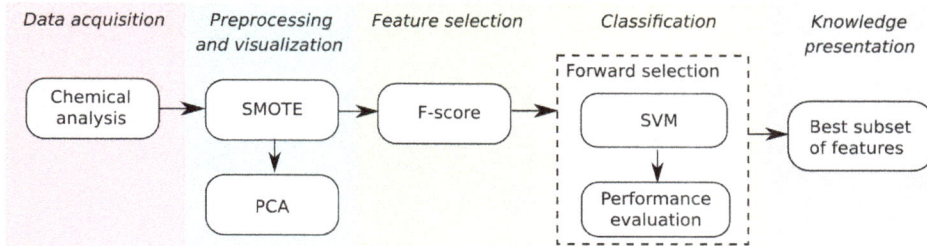

Figure 1. Schematic view of the Tannat wines analysis.

After that, we visualized data through the use of PCA, a descriptive tool that visualizes the data in two dimensions. The *F*-score, a filter feature selection method, were used to rank features in the order of importance. According to the feature order generated, we constructed 20 feature subsets through an iterative forward-selection procedure. The feature with the highest *F*-score value was assigned to the first subset (#1). The top two features on the *F*-score ranking were assigned to the second subset (#2), and so on, until all features were assigned to the twentieth subset (#20).

The support vector machines classifier, with the radial basis function kernel, was used to build a classification model for each feature subset. This included a grid search on the classifier parameters for the leave-one-out cross-validation with training, validation, and test sets. Finally, four performance measures were computed for each model, based on the test set.

The entire analysis was conducted using R software [34]. R is a free software environment for statistical computing and graphics that contains a wide variety of packages for performing statistical analysis. In this study, we implemented the *F*-score algorithm and used the caret package [35] for data classification, as well as the ggplot2 package [36] to visualize some of the results.

2.9. Support Vector Machines

The support vector machine (SVM) [37] is a supervised learning model that analyzes data used for classification and regression analysis, based on the statistical learning theory and structural risk minimization. The algorithm used in this model obtains an optimal hyperplane with a maximum margin for separating the classes of samples. The generalization power with which the hyperplane separates the classes depends on its margin, defined as the distance between the hyperplane and the samples closest to it (support vector). This classifier is one of the most robust and accurate methods among the well-known data mining algorithms [38]. Additionally, it is a useful classification algorithm when few training data are available. It is also a suitable classifier for balanced datasets [20].

Given a two-class training set whose classes are linearly separable, i.e., all training samples can be correctly classified by the hyperplane, the algorithm computes the decision boundary based on samples that are closest to the maximum-margin hyperplane, which are called support vectors. This hyperplane is represented by Equation (3), in which **w** is a weight vector, x is the input data and b is a bias:

$$\mathbf{w} \cdot x + b = 0 \tag{3}$$

The goal is to maximize the margin of this hyperplane, increasing the distance between the samples within its limit. SVM finds an optimum separating hyperplane that maximizes the margin of the decision surface, by solving the following Equation (4):

$$\min_{\mathbf{w},b} \tfrac{1}{2}\mathbf{w}^t\mathbf{w}$$
$$s.t.\ y_i(\mathbf{w}{\cdot}x_i + b) - 1 \geq 0 \tag{4}$$

However, in many real-world problems, it may not be possible to trace a decision boundary that separates the samples into the class labels. In these cases, it is necessary to apply a kernel function and to add slack variables. Thus, the SVM formulation becomes Equation (5):

$$\min_{\mathbf{w},b,\xi_i} \frac{\|\mathbf{w}^2\|}{2} + C\left(\sum_{i=1}^{n} \xi_i\right)$$
$$s.t.\ y_i(\mathbf{w}{\cdot}\phi(x_i) + b) \geq 1 - \xi_i,\ \ \xi_i \geq 0,\ i = 1,\ \cdots,\ n. \tag{5}$$

The kernel function Φ, based on the inner product between given data, performs a nonlinear transformation of data from the input space to a feature space with higher (even infinite) dimensions in order to make the problem linearly separable, as defined as Equation (6):

$$K(x_i,\ x_j) = \Phi(x_i){\cdot}\Phi(x_j) \tag{6}$$

In this study, we used the Gaussian radial basis function (RBF) kernel, defined by Equation (7):

$$K(x_i, x_j) = \exp\left(-\frac{\|\,x_i - x_j\,\|^2}{2\,\sigma^2}\right) \tag{7}$$

The C and σ parameters of Equations (5) and (7), respectively, are defined empirically, with a grid search taking place among the following values: $C = 2^c$, $c = \{-5, -3, \ldots, 5\}$, and $\sigma = 2^s$, $s = \{-10, -8, \ldots, 3\}$. Furthermore, RBF can produce a good performance, even when using a small number of samples [39].

2.10. Variable Selection

Variable selection methods provide a way of reducing computation time, improving prediction performance, and giving a better understanding of the data in machine learning applications [40]. To consider that there are possibly irrelevant or redundant variables in the data is important for improving the classification model. In this study, the F-score was employed to generate a ranking of importance [41]. The F-score is simple, generally quite effective, and has been used in previous studies [42–45]. Given training vectors x_k, $k = \{1, \ldots, m\}$ and the number of positive and negative instances n^+ and n^-, respectively, the F-score of the i-th feature is defined by Equation (8):

$$F_i = \frac{(\overline{x}_i^+ - x_i) + (\overline{x}_i^- - x_i)}{\frac{1}{n_+ - 1}\sum_{k=1}^{n^+}\left(\overline{x}_{k,i}^{(+)} - \overline{x}_i^{(+)}\right)^2 + \frac{1}{n_- - 1}\sum_{k=1}^{n^-}\left(\overline{x}_{k,i}^{(-)} - \overline{x}_i^{(-)}\right)^2}, \tag{8}$$

where $\overline{x}_i$, $\overline{x}_i^{(+)}$, and $\overline{x}_i^{(-)}$ are the average of the i-th feature of the whole, positive, and negative data sets, respectively; $x_{ki}^{(+)}$ is the i-th feature of the k-th positive instance, and $x_{ki}^{(-)}$ is the i-th feature of the k-th negative instance. The numerator indicates the discrimination between the positive and negative sets, and the denominator indicates the value within each of the two sets. The larger the F-score, the more likely that this feature will be more discriminative.

2.11. Performance Analysis

To evaluate the performance of a classification model, it is customary to split the data into two general parts, a training set and a test set, using holdout (70%–30%), 10-fold cross validation (k-fold CV), or leave-one-out cross validation (LOOCV) methodology. The holdout method is not indicated when the dataset has a small number of samples, making this method unfeasible in our study. Cross-validation is one possible solution for training and testing sets that have a small number of samples [32].

Due to our limited number of samples, a LOOCV strategy was employed to evaluate the performance of the classifier. This method is a particular case of the k-fold cross validation technique, which randomly splits data set D into k subsets $D_1, D_2, \ldots, D_k$ (the folds) of approximately equal size. In the LOOCV method, k is equal to the number of samples. The process of building the classification model occurs k times; the training set (k-1 folds) is used to perform the classification model, and the prediction ability was tested on the samples of the omitted fold. In the training phase, we performed an internal 10-fold cross-validation to validate the model and to select the best classifier parameters. This way, each sample was used to test the model without influencing the training phase, in order to avoid the introduction of bias.

After performing the classification models, it was possible to estimate the accuracy and percentage of correct predictions for each class of data. The results of the classification are established in a confusion matrix, as shown below in Table 1.

Table 1. Confusion matrix example.

Reference	Prediction	
	Positive	**Negative**
Positive	TP	FN
Negative	FP	TN

Interpreting the confusion matrix as presented with the data analyzed, we see that the positive samples (+) are from Brazil and the negative samples (−) are from Uruguay. The matrix values are true positive (TP) for samples correctly classified as positive, true negative (TN) for samples correctly classified as negative, false negative (FN) for the positive samples that were classified as negative, and false positive (FP) for negative samples that were classified as positive. From these values, we see that it is possible to establish the performance measures used in this study. The higher these parameters, the better the result:

$$\text{Accuracy} = \frac{TP + TN}{TP + TN + FP + FN} \times 100, \tag{9}$$

$$\text{Sensitivity} = \frac{TP}{TP + FN} \times 100, \tag{10}$$

$$\text{Specificity} = \frac{TN}{TN + FP} \times 100, \tag{11}$$

$$\text{MCC} = \frac{TP \times TN - FP \times FN}{\sqrt{(TP + FP)(TP + FN)(TN + FP)(TN + FN)}}. \tag{12}$$

The accuracy reflects how closely the classifier reaches its goal, that is, the accuracy mirrors the percentage of the model that has been correct in its predictions. Sensitivity refers to the percentage of correct answers with regard to the positive samples. Specificity is the opposite of sensitivity. It measures the percentage of samples correctly classified as negative. The Matthew's correlation coefficient (MCC) is, in principle, a correlation coefficient that lies between the real class and the predicted class for binary classifications. The MCC returns a value between −1 and +1, where +1 represents a perfect prediction, −1 indicates total disagreement between prediction and the reference, and 0 means no better than random prediction.

3. Results and Discussion

The methodology starts with the creation of new synthetic data, followed by the application of the feature selection method F-score on all features in order to generate a ranking of importance. Table 2 describes the interval of concentrations for each variable. An exploratory PCA shown in Figure 1 reveals that the two classes of wines cannot be differentiated. The PC1–PC2 scores' subspace accounted for 100% of the original variance. Figure 2a shows the original data and Figure 2b reveals the dataset resulting from the combination of the original data and 19 new synthetic samples. The new samples occupied the same region on the PCA plot as the original samples.

Table 2. Values of mean, standard deviation, and minimum and maximum variable concentration for each country.

Variable	Brazil (*n* = 28)	Uruguay (*n* = 28)
L *	16.02 ± 5.91 (3.84–31.02)	17.48 ± 6.25 (4.7–32.43)
a *	45.47 ± 6.74 (25.7–54.17)	47.24 ± 5.84 (29.43–54.51)
b *	26.35 ± 8.63 (6.37–42.87)	28.95 ± 9.59 (7.61–48.6)
TPI	1924.45 ± 352.61 (1365.36–2999.55)	1973.45 ± 454.56 (1015.98–2946.75)
TA	143.41 ± 35.32 (95.35–233.45)	117.02 ± 52.93 (28.72–289.56)
ORAC	39,623.59 ± 6942.55 (25,423.23–56,914.9)	44,431.93 ± 9938.45 (26,883.63–69,192.46)
DPPH	8.88 ± 0.58 (7.55–9.82)	9.56 ± 0.67 (7.12–10.34)
cyan-3-glu	0.22 ± 0.12 (0.1–0.62)	0.18 ± 0.07 (0.07–0.36)
delph-3-acetylglu	1.15 ± 0.81 (0.48–3.61)	0.98 ± 0.7 (0.33–3.83)
delph-3-glu	6.19 ± 2.71 (2.17–11.38)	3.65 ± 2.63 (0.29–11.4)
malv-3-(coum)glu	6.36 ± 2.89 (2.97–11.84)	5.04 ± 3.53 (0.41–14.18)
malv-3-acetylglu	13.05 ± 5.69 (4.55–27.78)	11.01 ± 7.35 (1.04–30.79)
malv-3-glu	42.58 ± 22.03 (26.03–95.19)	36.89 ± 22.79 (3.86–89.98)
peon-3-(coum)glu	1.48 ± 1.98 (0.57–4.71)	1.76 ± 2.07 (0.27–11.11)
peon-3-acetylglu	1.57 ± 0.74 (0.89–3.25)	1.41 ± 0.61 (0.51–2.9)
peon-3-glu	4.27 ± 2.12 (1.61–9.99)	2.11 ± 1.46 (0.4–5.72)
pet-3-(coum)glu	0.45 ± 0.56 (0.1–2.14)	0.28 ± 0.29 (0.1–1.64)
pet-3-acetylglu	2.43 ± 2.02 (1.04–8.63)	1.57 ± 1.74 (0–9.1)
pet-3-glu	11.02 ± 3.97 (4.98–22.14)	7.16 ± 5.21 (0.46–24.52)
vitisin A	14.37 ± 5.2 (5.37–22.25)	11.32 ± 8.89 (1.95–43.75)

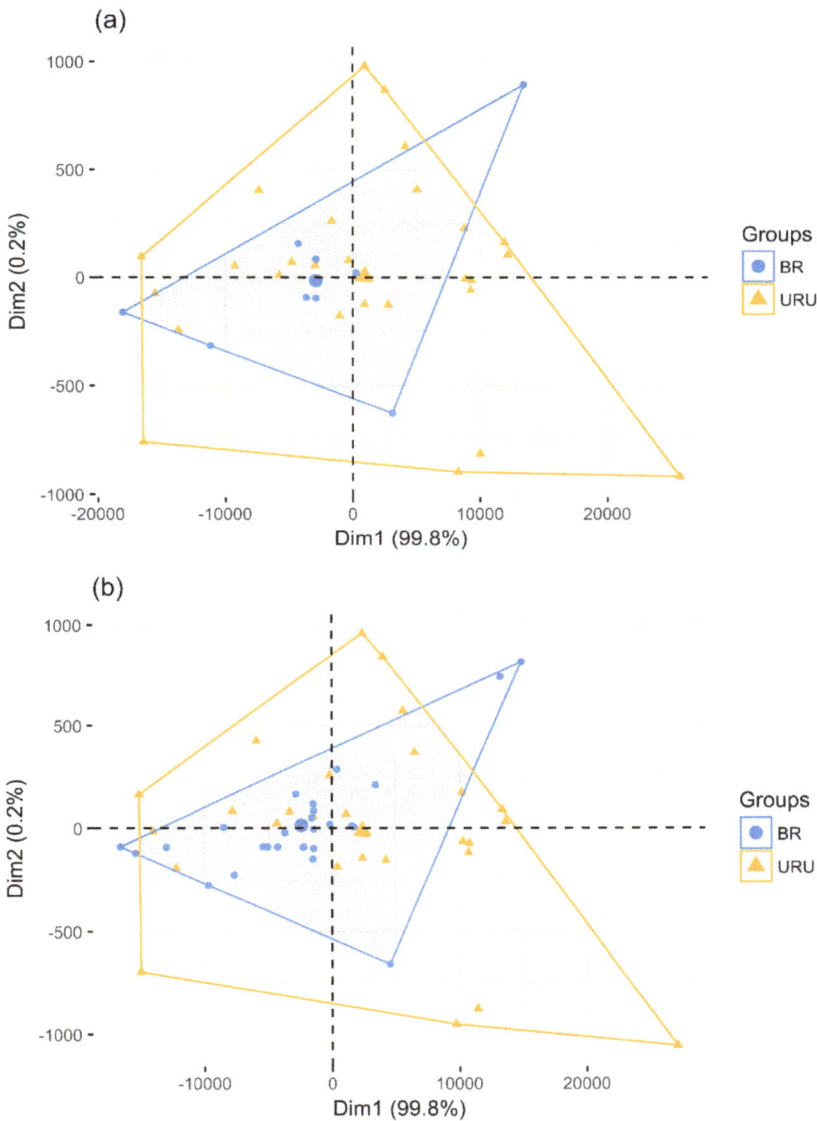

Figure 2. Visual separation of wines based on PC1 (principal component) and PC2. (**a**) The original data; (**b**) the synthetic data added to the original data.

3.1. Classification Analysis

At first, we classified only the original samples in order to obtain a reference result for comparing this outcome with the balanced dataset results. The classification model with all features along with LOOCV resulted in the following performance measures: 72.97% of accuracy, an MCC of 0.45, a sensitivity of 40%, and 78.12% of specificity. These results indicate that the few Brazilian samples could be poorly classified, and that the model benefited the majority class. In addition, the MCC metric indicates that the predictions were closer to being random than perfect predictions.

The next step was to generate the ranking of importance and perform the classification models. Figure 3 shows the *F*-score ranking of importance. Then, we constructed 20 feature subsets, an iterative forward-selection procedure based on the *F*-score ranking of importance. The first feature subset (#1) is composed of the peon-3-glu compound; the second feature subset (#2) is composed of the peon-3-glu and DPPH parameters, and so forth; until the twentieth feature subset (#20), which is composed of all the parameters. Table 3 presents the results obtained from the classification models of the balanced data.

Table 3. Overall results from leave-one-out cross validation (LOOCV) classification with the support vector machine (SVM) for each feature subset.

#	Compounds	Acc	MCC	Sens	Spec
#1	peon-3-glu	91.07	0.82	89.66	92.59
#2	peon-3-glu, DPPH	94.64	0.90	90.32	100
#3	peon-3-glu, DPPH, delph-3-glu	85.71	0.71	85.71	85.71
#4	peon-3-glu, DPPH, delph-3-glu, pet-3-glu	92.86	0.86	90.00	96.15
#5	peon-3-glu, DPPH, delph-3-glu, pet-3-glu, TA	91.07	0.82	89.66	92.59
#6	peon-3-glu, DPPH, delph-3-glu, pet-3-glu, TA, ORAC	94.64	0.90	90.32	100
#7	peon-3-glu, DPPH, delph-3-glu, pet-3-glu, TA, ORAC, pet-3-acetylglu	91.07	0.83	87.10	96.00
#8	peon-3-glu, DPPH, delph-3-glu, pet-3-glu, TA, ORAC, pet-3-acetylglu, vitisin A	94.64	0.90	90.32	100
#9	peon-3-glu, DPPH, delph-3-glu, pet-3-glu, TA, ORAC, pet-3-acetylglu, vitisin A, cyan-3-glu	91.07	0.83	87.10	96.00
#10	peon-3-glu, DPPH, delph-3-glu, pet-3-glu, TA, ORAC, pet-3-acetylglu, vitisin A, cyan-3-glu, malv-3-(coum)glu	91.07	0.83	87.10	96.00
#11	peon-3-glu, DPPH, delph-3-glu, pet-3-glu, TA, ORAC, pet-3-acetylglu, vitisin A, cyan-3-glu, malv-3-(coum)glu, pet-3-(coum)glu	85.71	0.72	83.33	88.46
#12	peon-3-glu, DPPH, delph-3-glu, pet-3-glu, TA, ORAC, pet-3-acetylglu, vitisin A, cyan-3-glu, malv-3-(coum)glu, pet-3-(coum)glu, malv-3-acetylglu	92.86	0.87	87.50	100
#13	peon-3-glu, DPPH, delph-3-glu, pet-3-glu, TA, ORAC, pet-3-acetylglu, vitisin A, cyan-3-glu, malv-3-(coum)glu, pet-3-(coum)glu, malv-3-acetylglu, b *	89.29	0.79	86.67	92.31
#14	peon-3-glu, DPPH, delph-3-glu, pet-3-glu, TA, ORAC, pet-3-acetylglu, vitisin A, cyan-3-glu, malv-3-(coum)glu, pet-3-(coum)glu, malv-3-acetylglu, b *, a *	91.07	0.83	87.10	96.00
#15	peon-3-glu, DPPH, delph-3-glu, pet-3-glu, TA, ORAC, pet-3-acetylglu, vitisin A, cyan-3-glu, malv-3-(coum)glu, pet-3-(coum)glu, malv-3-acetylglu, b *, a *, malv-3-glu	89.29	0.79	86.67	92.31
#16	peon-3-glu, DPPH, delph-3-glu, pet-3-glu, TA, ORAC, pet-3-acetylglu, vitisin A, cyan-3-glu, malv-3-(coum)glu, pet-3-(coum)glu, malv-3-acetylglu, b *, a *, malv-3-glu, L *	92.86	0.87	87.50	100
#17	peon-3-glu, DPPH, delph-3-glu, pet-3-glu, TA, ORAC, pet-3-acetylglu, vitisin A, cyan-3-glu, malv-3-(coum)glu, pet-3-(coum)glu, malv-3-acetylglu, b *, a *, malv-3-glu, L *, peon-3-acetylglu	91.07	0.83	87.10	96.00
#18	peon-3-glu, DPPH, delph-3-glu, pet-3-glu, TA, ORAC pet-3-acetylglu, vitisin A, cyan-3-glu, malv-3-(coum)glu, pet-3-(coum)glu, malv-3-acetylglu, b *, a *, malv-3-glu, L *, peon-3-acetylglu, delph-3-acetylglu	91.07	0.83	87.10	96.00
#19	peon-3-glu, DPPH, delph-3-glu, pet-3-glu, TA, ORAC, pet-3-acetylglu, vitisin A, cyan-3-glu, malv-3-(coum)glu, pet-3-(coum)glu, malv-3-acetylglu, b *, a *, malv-3-glu, L *, peon-3-acetylglu, delph-3-acetylglu, peon-3-(coum)glu	92.86	0.87	87.50	100
#20	peon-3-glu, DPPH, delph-3-glu, pet-3-glu, TA, ORAC, pet-3-acetylglu, vitisin A, cyan-3-glu, malv-3-(coum)glu, pet-3-(coum)glu, malv-3-acetylglu, b *, a *, malv-3-glu, L *, peon-3-acetylglu, delph-3-acetylglu, peon-3-(coum)glu, TPI	92.86	0.87	87.50	100

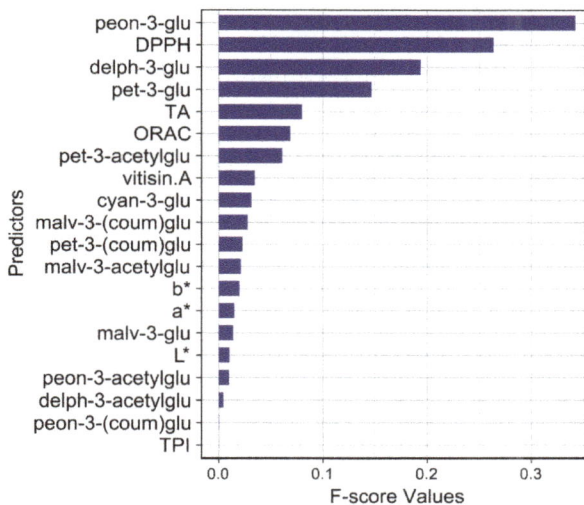

Figure 3. *F*-score ranking of importance.

The best overall performances occurred with the use of at least two, six, or eight variables. The classification models built with datasets #2, #6, and #8 resulted in an accuracy of 94.64% and an MCC of 0.90; the Brazilian samples were correctly classified at a rate of 90.32% (sensitivity) and all Uruguayan samples were correctly classified. The performances differ for each combination of input variables. The main parameters, which resulted in almost 95% of accuracy, included peon-3-glu, DPPH, delph-3-glu, pet-3-glu, TA, ORAC, pet-3-acetylglu, and vitisin A. It was interesting to note that through the use of only one compound (peon-3-glu), the classifier performance is very good, achieving 91% of accuracy and an MCC of 0.82. An equally interesting result is that even with the use of all compounds, the accuracy improved only slightly, to 92.86%.

It is possible to observe how the classification model was improved by the addition of synthetic Brazilian samples when comparing the result of the first classification model performed in this study (containing only the original samples) to the classification performed with a new dataset composed of the original and synthetic samples. Using all twenty compounds of the balanced dataset as input data, the model achieved an accuracy of 92.86%, an MCC of 0.87, and a sensitivity and specificity of 87.5% and 100%, respectively. The performance obtained from the use of only the original data was lower, however, with poor and random classifications (72.97% of accuracy, an MCC of 0.45, a sensitivity of 40%, and 78.12% of specificity). Therefore, it is notable that both class predictions, Brazilian and Uruguayan, were improved.

3.2. Analysis of Variable Importance

The group of anthocyanidin-3-glucosides comprises the main group of anthocyanins in wine, with malvidin-3-glucoside usually being the main component. In our study, three out of the five individual anthocyanins that were influential with regard to the classification of Tannat wines' (peon-3-glu, delph-3-glu, and pet-3-glu) are part of this group. In spite of being in the majority, malv-3-glu is not one of the main variables responsible for the classification. Gutiérrez et al. found that malv-3-glu did not differentiate Chilean Merlot, Carménère, and Cabernet Sauvignon wines that were produced in different valleys [46]. On the other hand, they found that malv-3-(coum)glu played an important role, together with other anthocyanins, in this differentiation. In a previous study of the Cabernet Sauvignon classification according to geographical origin (Brazil and Chile), we found that L *,

DPPH, delph-3-acetylglu, peon-3-(coum)glu, and pet-3-acetylglu were relevant variables for the classification [25].

Tannat wines have lower methyl–transferase activity compared to other wine varieties [7]. This allows the accumulation of delphinidins and petunidins, making these anthocyanins especially important in Tannat. This could explain, at least in part, the important role of delph-3-glu and pet-3-glu in Tannat wine classification. Another possible contribution pertains to the percentage of primitive anthocyanins (delph-3-glu and pet-3-glu), which may provide a significant indicator of the weather conditions pertaining to the ripening of grapes [47]. Therefore, these two anthocyanins could contribute to wine classification due to climatic differences in different wine production regions, which possibly affect the ripening process.

We then separately analyzed the most discriminating parameters according to the country of origin. Figure 4 presents the boxplot for each compound, stratified according to the country or origin. We note that most of the parameter values overlap. Most of the anthocyanin concentrations (including TA) are lower for the Uruguayan samples, while the DPPH and ORAC values of these samples are slightly higher, indicating that the anthocyanins are not the most relevant contributors to antioxidant activity.

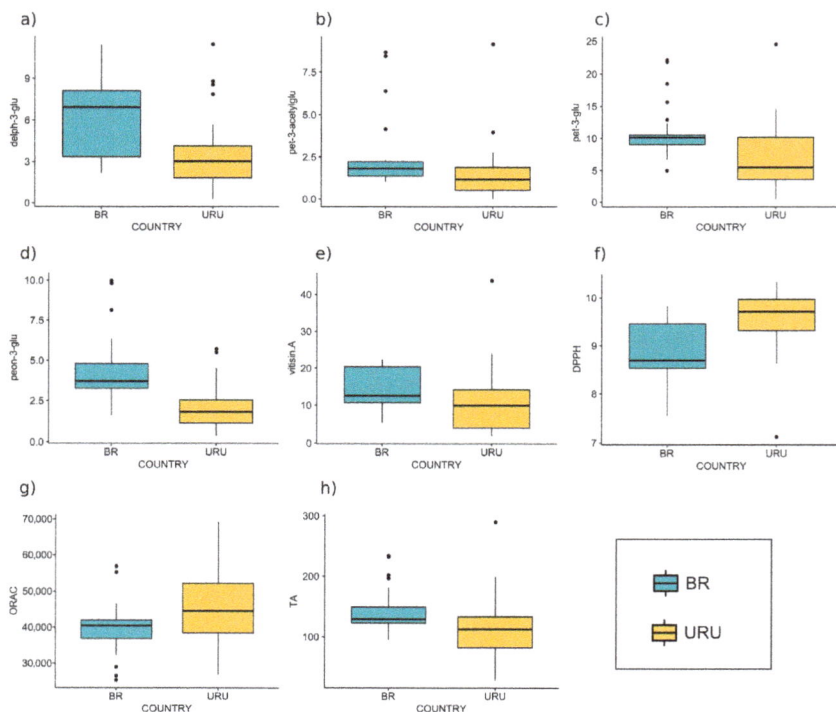

Figure 4. Boxplots of the eight compounds selected by the *F*-score, along with SVM classification. (**a**) delph-3-glu variable; (**b**) pet-3-acetylglu variable; (**c**) pet-3-glu variable; (**d**) peon-3-glu; (**e**) vitisin A variable; (**f**) free radical scavenging capacity (DPPH) variable; (**g**) oxygen radical absorbance capacity (ORAC) variable; and (**h**) total anthocyanins (TA) variable.

4. Conclusions

In this study, we applied data mining techniques to classify Brazilian and Uruguayan Tannat wine samples. To our knowledge, this paper is the first to classify Tannat wines using SVM and

Beverages **2018**, *4*, 97

to identify the variables that contributed most significantly to the classification. Using the original imbalanced dataset (28 Uruguayan samples and 9 Brazilian samples), the classification model achieved an accuracy rate of 72.97% and an MCC of 0.45. This performance improved when we added new synthetic Brazilian samples to the dataset. The best overall accuracy (94.64) and MCC (0.90) occurred when at least two, six, or eight variables were used, including, in the following order: peon-3-glu, DPPH, delph-3-glu, pet-3-glu, TA, ORAC, pet-3-acetylglu, and vitisin A. This finding proves that these variables are the most relevant for differentiating Tannat wine samples with respect to their geographical origin, and that the addition of synthetic samples can improve the classification model.

The identification of the most important parameters is useful for reducing time and resources put forth when classifying future wines. Additionally, understanding the behavior of wine chemical components is also important, as that knowledge serves as an information source for companies to preserve, maintain, and ensure the quality of wine, and to avoid fraud. Furthermore, the soil, climatic conditions, type of harvest, and conditions for production contribute to the characteristics that make a unique wine from a particular region, which explains the difference found among the parameter concentrations for each respective country in this study. The use of SVM allowed researchers to discriminate between the regions, whereas the PCA score plot only shows the overlapping of samples for each class. Therefore, this methodology can be applied for certification purposes that, in general, involved the origin of other wines and food products.

Author Contributions: Conceptualization, R.B. and I.A.C.; Methodology, N.L.C.; Chemical Analysis, I.A.C. and L.A.G.L.; Data Analysis, N.L.C.; Investigation, N.L.C.; Data Curation, N.L.C., I.A.C, and L.A.G.L.; Writing-Original Draft Preparation, N.L.C. and L.A.G.L; Writing-Review and Editing, R.B. and L.A.G.L.; Supervision, R.B.

Funding: This study was financed in part by the Coordenação de Aperfeiçoamento de Pessoal de Nível Superior—Brasil (CAPES)—Finance Code 001.

Conflicts of Interest: The authors declare no conflict of interest.

References

1. Gómez-Meire, S.; Campos, C.; Falqué, E.; Díaz, F.; Fdez-Riverola, F. Assuring the authenticity of northwest Spain white wine varieties using machine learning techniques. *Food Res. Int.* **2014**, *60*, 230–240. [CrossRef]
2. Luykx, D.M.A.M.; van Ruth, S.M. An overview of analytical methods for determining the geographical origin of food products. *Food Chem.* **2008**, *107*, 897–911. [CrossRef]
3. Geana, E.I.; Popescu, R.; Costinel, D.; Dinca, O.R.; Stefanescu, I.; Ionete, R.E.; Bala, C. Verifying the red wines adulteration through isotopic and chromatographic investigations coupled with multivariate statistic interpretation of the data. *Food Control* **2016**, *62*, 1–9. [CrossRef]
4. Versari, A.; Laurie, V.F.; Ricci, A.; Laghi, L.; Parpinello, G.P. Progress in authentication, typification and traceability of grapes and wines by chemometric approaches. *Food Res. Int.* **2014**, *60*, 2–18. [CrossRef]
5. Juan Gennari, A. A letter by the Regional Editor for South America: From varietals to terroir. *Wine Econ. Policy* **2014**, *3*, 69–70. [CrossRef]
6. González-Neves, G.; Gil, G.; Barreiro, L.; Favre, G. Pigment profile of red wines cv. Tannat made with alternative winemaking techniques. *J. Food Compos. Anal.* **2010**, *23*, 447–454. [CrossRef]
7. González-Neves, G.; Franco, J.; Barreiro, L.; Gil, G.; Moutounet, M.; Carbonneau, A. Varietal differentiation of Tannat, Cabernet-Sauvignon and Merlot grapes and wines according to their anthocyanic composition. *Eur. Food Res. Technol.* **2007**, *225*, 111–117. [CrossRef]
8. Welke, J.E.; Manfroi, V.; Zanus, M.; Lazzarotto, M.; Zini, C.A. Differentiation of wines according to grape variety using multivariate analysis of comprehensive two-dimensional gas chromatography with time-of-flight mass spectrometric detection data. *Food Chem.* **2013**, *141*, 3897–3905. [CrossRef] [PubMed]
9. Xiao, Z.; Liu, S.; Gu, Y.; Xu, N.; Shang, Y.; Zhu, J. Discrimination of cherry wines based on their sensory properties and aromatic fingerprinting using HS-SPME-GC-MS and multivariate analysis. *J. Food Sci.* **2014**, *79*, C284–C294. [CrossRef] [PubMed]
10. Rešetar, D.; Marchetti-Deschmann, M.; Allmaier, G.; Katalinić, J.P.; Pavelić, S.K. Matrix assisted laser desorption ionization mass spectrometry linear time-of-flight method for white wine fingerprinting and classification. *Food Control* **2016**, *64*, 157–164. [CrossRef]

11. Płotka-Wasylka, J.; Simeonov, V.; Morrison, C.; Namieśnik, J. Impact of selected parameters of the fermentation process of wine and wine itself on the biogenic amines content: Evaluation by application of chemometric tools. *Microchem. J.* **2018**, *142*, 187–194. [CrossRef]

12. Bonello, F.; Cravero, M.; Dell'Oro, V.; Tsolakis, C.; Ciambotti, A. Wine Traceability Using Chemical Analysis, Isotopic Parameters, and Sensory Profiles. *Beverages* **2018**, *4*, 54. [CrossRef]

13. Aceto, M.; Bonello, F.; Musso, D.; Tsolakis, C.; Cassino, C.; Osella, D. Wine Traceability with Rare Earth Elements. *Beverages* **2018**, *4*, 23. [CrossRef]

14. Zielinski, A.A.F.; Haminiuk, C.W.I.; Nunes, C.A.; Schnitzler, E.; Ruth, S.M.; Granato, D. Chemical composition, sensory properties, provenance, and bioactivity of fruit juices as assessed by chemometrics: A critical review and guideline. *Compre. Rev. Food Sci. food Saf.* **2014**, *13*, 300–316. [CrossRef]

15. Callao, M.P.; Ruisánchez, I. An overview of multivariate qualitative methods for food fraud detection. *Food Control* **2018**, *86*, 283–293. [CrossRef]

16. Condurso, C.; Cincotta, F.; Tripodi, G.; Verzera, A. Characterization and ageing monitoring of Marsala dessert wines by a rapid FTIR-ATR method coupled with multivariate analysis. *Eur. Food Res. Technol.* **2018**, *244*, 1073–1081. [CrossRef]

17. Ríos-Reina, R.; García-González, D.L.; Callejón, R.M.; Amigo, J.M. NIR spectroscopy and chemometrics for the typification of Spanish wine vinegars with a protected designation of origin. *Food Control* **2018**, *89*, 108–116. [CrossRef]

18. Ioannou-Papayianni, E.; Kokkinofta, R.I.; Theocharis, C.R. Authenticity of Cypriot sweet wine commandaria using FT-IR and chemometrics. *J. Food Sci.* **2011**, *76*, C420–C427. [CrossRef] [PubMed]

19. Ropodi, A.I.; Panagou, E.Z.; Nychas, G.-J. Data mining derived from food analyses using non-invasive/non-destructive analytical techniques; determination of food authenticity, quality & safety in tandem with computer science disciplines. *Trends Food Sci. Technol.* **2016**, *50*, 11–25. [CrossRef]

20. Guo, H.; Li, Y.; Jennifer, S.; Gu, M.; Huang, Y.; Gong, B. Learning from class-imbalanced data: Review of methods and applications. *Expert Syst. Appl.* **2017**, *73*, 220–239. [CrossRef]

21. Majchrzak, T.; Wojnowski, W.; Płotka-Wasylka, J. Classification of Polish wines by application of ultra-fast gas chromatography. *Eur. Food Res. Technol.* **2018**, 1–9. [CrossRef]

22. Capron, X.; Massart, D.L.; Smeyers-Verbeke, J. Multivariate authentication of the geographical origin of wines: A kernel SVM approach. *Eur. Food Res. Technol.* **2007**, *225*, 559–568. [CrossRef]

23. Ríos-Reina, R.; Elcoroaristizabal, S.; Ocaña-González, J.A.; García-González, D.L.; Amigo, J.M.; Callejón, R.M. Characterization and authentication of Spanish PDO wine vinegars using multidimensional fluorescence and chemometrics. *Food Chem.* **2017**, *230*, 108–116. [CrossRef] [PubMed]

24. Jurado, J.M.; Alcázar, Á.; Palacios-Morillo, A.; De Pablos, F. Classification of Spanish DO white wines according to their elemental profile by means of support vector machines. *Food Chem.* **2012**, *135*, 898–903. [CrossRef] [PubMed]

25. da Costa, N.L.; Castro, I.A.; Barbosa, R. Classification of Cabernet Sauvignon from Two Different Countries in South America by Chemical Compounds and Support Vector Machines. *Appl. Artif. Intell.* **2016**, *30*, 679–689. [CrossRef]

26. Soares, F.; Anzanello, M.J.; Fogliatto, F.S.; Marcelo, M.C.A.; Ferrão, M.F.; Manfroi, V.; Pozebon, D. Element selection and concentration analysis for classifying South America wine samples according to the country of origin. *Comput. Electron. Agric.* **2018**, *150*, 33–40. [CrossRef]

27. Singleton, V.L.; Rossi, J.A. Colorimetry of total phenolics with phosphomolybdic-phosphotungstic acid reagents. *Am. J. Enol. Vitic.* **1965**, *16*, 144–158.

28. Fuleki, T.; Francis, F.J. Determination of total anthocyanin and degradation index for cranberry juice. *Food Sci.* **1968**, *33*, 78–83. [CrossRef]

29. Boido, E.; Alcalde-Eon, C.; Carrau, F.; Dellacassa, E.; Rivas-Gonzalo, J.C. Aging effect on the pigment composition and color of *Vitis vinifera* L. cv. Tannat wines. Contribution of the main pigment families to wine color. *J. Agric. Food Chem.* **2006**, *54*, 6692–6704. [CrossRef] [PubMed]

30. Arnous, A.; Makris, D.P.; Kefalas, P. Effect of principal polyphenolic components in relation to antioxidant characteristics of aged red wines. *J. Agric. Food Chem.* **2001**, *49*, 5736–5742. [CrossRef] [PubMed]

31. Huang, D.; Ou, B.; Hampsch-Woodill, M.; Flanagan, J.A.; Prior, R.L. High-throughput assay of oxygen radical absorbance capacity (ORAC) using a multichannel liquid handling system coupled with a microplate fluorescence reader in 96-well format. *J. Agric. Food Chem.* **2002**, *50*, 4437–4444. [CrossRef] [PubMed]

Beverages **2018**, *4*, 97

32. Witten, I.H.; Frank, E.; Hall, M.A. *Data Mining: Practical Machine Learning Tools and Techniques*; Elsevier: Amsterdam, The Netherlands, 2011.

33. Chawla, N.V.; Bowyer, K.W.; Hall, L.O.; Kegelmeyer, W.P. SMOTE: Synthetic minority over-sampling technique. *J. Artif. Intell. Res.* **2002**, *16*, 321–357. [CrossRef]

34. Team, the R.C. R: A language and environment for statistical computing. R Foundation for Statistical Computing. Available online: https://www.R-project.org/ (accessed on 30 November 2018).

35. Kuhn, M. The caret package. Available online: http://caret.r-forge.r-project.org/ (accessed on 30 November 2018).

36. Wickham, H. *Ggplot2: Elegant Graphics for Data Analysis*; Springer-Verlag New York: New York, NY, USA, 2009.

37. Cortes, C.; Vapnik, V. Support-vector networks. *Mach. Learn.* **1995**, *20*, 273–297. [CrossRef]

38. Xue, H.; Yang, Q.; Chen, S. SVM: Support vector machines. In *The Top Ten Algorithms in Data Mining*; Taylor & Francis Group: Abingdon-on-Thames, UK, 2009; pp. 37–59.

39. Noori, R.; Abdoli, M.A.; Ghasrodashti, A.A.; Jalili Ghazizade, M. Prediction of municipal solid waste generation with combination of support vector machine and principal component analysis: A case study of Mashhad. *Environ. Prog. Sustain. Energy* **2009**, *28*, 249–258. [CrossRef]

40. Chandrashekar, G.; Sahin, F. A survey on feature selection methods. *Comput. Electr. Eng.* **2014**, *40*, 16–28. [CrossRef]

41. Chen, Y.-W.; Lin, C.-J. Combining SVMs with various feature selection strategies. In *Feature Extraction*; Springer: Berlin, Germany, 2006; pp. 315–324.

42. Liu, F.; Guo, W.; Fouche, J.-P.; Wang, Y.; Wang, W.; Ding, J.; Zeng, L.; Qiu, C.; Gong, Q.; Zhang, W.; et al. Multivariate classification of social anxiety disorder using whole brain functional connectivity. *Brain Struct. Funct.* **2015**, *220*, 101–115. [CrossRef] [PubMed]

43. Turra, C.; de Lima, M.D.; Fernandes, E.A.D.N.; Bacchi, M.A.; Barbosa, F.; Barbosa, R. Multielement determination in orange juice by ICP-MS associated with data mining for the classification of organic samples. *Inf. Process. Agric.* **2017**, *4*, 199–205. [CrossRef]

44. Marin, G.; Dominio, F.; Zanuttigh, P. Hand gesture recognition with jointly calibrated leap motion and depth sensor. *Multimed. Tools Appl.* **2016**, *75*, 14991–15015. [CrossRef]

45. Adnane, M.; Belouchrani, A. Heartbeats classification using QRS and T waves autoregressive features and RR interval features. *Expert Syst.* **2017**, *34*, e12219. [CrossRef]

46. Gutiérrez, L.; Quintana, F.A.; Baer, D.; von Mardones, C. Multivariate Bayesian discrimination for varietal authentication of Chilean red wine. *J. Appl.* **2011**, *4763*, 1–28. [CrossRef]

47. Río Segade, S.; Orriols, I.; Gerbi, V.; Rolle, L. Phenolic characterization of thirteen red grape cultivars from galicia by anthocyanin profile and flavanol composition. *J. Int. Sci. Vigne Vin* **2009**, *43*, 189–198. [CrossRef]

beverages

MDPI

Article

Mineral Composition through Soil-Wine System of Portuguese Vineyards and Its Potential for Wine Traceability

Sofia Catarino [1,2,3,*], Manuel Madeira [4], Fernando Monteiro [4], Ilda Caldeira [2,5], Raúl Bruno de Sousa [1] and António Curvelo-Garcia [2]

[1] LEAF—Linking Landscape, Environment, Agriculture and Food, Instituto Superior de Agronomia, Universidade de Lisboa, Tapada da Ajuda, 1349-017 Lisboa, Portugal; brunosousa@isa.ulisboa.pt
[2] INIAV—Instituto Nacional de Investigação Agrária e Veterinária, 2565-191 Dois Portos, Portugal; ilda.caldeira@iniav.pt (I.C.); ascurvelogarcia@gmail.com (A.C.-G.)
[3] CEFEMA—Center of Physics and Engineering of Advanced Materials, Instituto Superior Técnico, Universidade de Lisboa, Av. Rovisco Pais, 1, 1049-001 Lisboa, Portugal
[4] CEF—Forest Research Centre, Instituto Superior de Agronomia, Universidade de Lisboa, Tapada da Ajuda, 1349-017 Lisboa, Portugal; mavmadeira@isa.ulisboa.pt (M.M.); fgmonteiro@isa.ulisboa.pt (F.M.)
[5] ICAAM, Universidade de Évora, Pólo da Mitra, Ap. 94, 7002-554 Évora, Portugal
[*] Correspondence: sofiacatarino@isa.ulisboa.pt; Tel.: +35-121-365-3246

Received: 29 September 2018; Accepted: 7 November 2018; Published: 9 November 2018

Abstract: The control of geographic origin is one of a highest priority issue regarding traceability and wine authenticity. The current study aimed to examine whether elemental composition can be used for the discrimination of wines according to geographical origin, taking into account the effects of soil, winemaking process, and year of production. The elemental composition of soils, grapes, musts, and wines from three DO (Designations of Origin) and for two vintage years was determined by using the ICP-MS semi-quantitative method, followed by multivariate statistical analysis. The elemental composition of soils varied according to geological formations, and for some elements, the variation due to soil provenance was also observed in musts and wines. Li, Mn, Sr and rare-earth elements (REE) allowed wine discrimination according to vineyard. Results evidenced the influence of winemaking processes and of vintage year on the wine's elemental composition. The mineral composition pattern is transferred through the soil-wine system, and differences observed for soils are reflected in grape musts and wines, but not for all elements. Results suggest that winemaking processes and vintage year should be taken into account for the use of elemental composition as a tool for wine traceability. Therefore, understanding the evolution of mineral pattern composition from soil to wine, and how it is influenced by the climatic year, is indispensable for traceability purposes.

Keywords: geographic origin; geological material; multi-element composition; rare earth elements; vinification

1. Introduction

Closely linked to the perception that terroir determines the quality and character of wines, the control of geographic origin is one of the most challenging and highest priority issues regarding traceability and wine authenticity [1]. Regional differences in sensory characteristics of wines have commonly been attributed to, among others things, (e.g., grape variety, technological processes, and vintage), local variations in soil composition [2]. Over the past two decades, many efforts have been made to identify potential fingerprints and develop reliable analytical methods to determine the authenticity of wine [3–9]. Soil-related fingerprints justify special attention, given that there is a relationship between the chemical composition of wine and the composition of the provenance soil.

The most explored fingerprinting techniques combine chemical analysis, namely elemental and isotope ratio analysis, and multivariate statistical analysis of the chemical data to classify wines according to their geographical origin. The successful application of the techniques based on the multi-element composition of a wine strongly depends on the selection of suitable elements that would reflect the relationship with soil geochemistry, and therefore, have discriminating potential. Thus, the data on mineral elements in wine as a probe for origin determination has to be carefully interpreted, since there are many environmental, agricultural, and oenological factors that can easily mask vital elemental information [10,11]. Otherwise, wine classification will reflect not only the geographical provenance, but also anthropogenic factors.

An important source of metal content in wine comes from the vineyard soil via grapevine roots, being influenced by soil geochemistry and vine rootstock, among other factors. Other potential sources, introduced during the processing stages, from vine culture to aged wine, are atmospheric pollution, soil amendments, fertilizers, pesticides, irrigation water, contact materials during transport, vinification, and aging processes, enological processing aids, and additives [5,12–14]. Also, depletion of some elements occurs over time, especially during alcoholic fermentation. Precipitation of K and Ca as tartrate salts begins during alcoholic fermentation and continues during the aging period. The precipitation of heavy metals as insoluble salts, namely as sulfides, is favoured by sulfur dioxide addition during winemaking [2]. All these factors may markedly change the multi-element composition of the wine, affecting the relationship between wine and soil compositions, thus precluding their use for authentication purposes.

Alkaline elements, Li, Rb, and Cs, are good indicators of geographical origin, as they are not included in the group of contaminant elements of wines [10]. Mn, Mg, Sr, Ba, and rare earth elements (REE) are also listed as useful elements, although the first two should be considered carefully, as they can be introduced through viticultural practices such as the use of fertilizers and pesticides, while REE content in wines can increase due to treatment with bentonites [11].

Studies on this subject have been pursued in most wine producing-countries, such as Argentina [15], Australia [7,16], Brazil [17], Canada [3,18], Germany [5], Italy [19,20], Portugal [4,21], Romania [22,23], South Africa [8,24,25], and Spain [26,27], indicating that the multi-elemental determination of wines can enable their successful discrimination. Most of the aforementioned studies are exclusively focused on wines produced under controlled conditions in order to guarantee their authenticity. Studies, involving wines and the provenance soils, are scarce [3,4,22].

The simultaneous use of a greater number of variables can provide increasingly robust results for the identification of wine geographical origin. Some studies based on this approach are described, with the most chosen variables being multi-element composition, strontium isotopic ratio ($^{87}Sr/^{86}Sr$), and stable isotope ratios of light elements [6,28]. Within a research program regarding strategies for wine fingerprinting, REE and $^{87}Sr/^{86}Sr$ were identified as viable tools for traceability of Portuguese DO in wines, where soils are developed on different geological formations [6,9], and studies on their robustness are underway [29,30].

Portugal is the eleventh largest producer of wine and the nine largest exporter in the world, with over six hundred millions of liters produced in 2016 [31]. Nevertheless, there is little information on multi-elemental analysis of Portuguese wines for their classification according to geographical origin. In this context, a study was developed to investigate whether inter-regional variation in multi-element composition could be used as a tool for the traceability of three Portuguese Designations of Origin (DO), where soils are developed on different geological formations. Having in mind the natural heterogeneity of some wine regions in terms of geological materials and soils, intra-regional variability was also studied by involving different vineyards from the same DO. Also, the evolution of mineral composition during vinification and inter-year variability were considered. Results will enlarge global databanks on wine composition and support comparisons with other world regions.

2. Materials and Methods

2.1. Vineyards

Four vineyards from three Portuguese DO (Dão, Óbidos and Palmela) were studied. In the Palmela DO (Southern Portugal), one vineyard of José Maria da Fonseca was considered: Vinha de Algeruz (AL; 38°34′ N, 8°49′ W), established on Haplic Eutric Arenosols, and Regosols, developed on Pliocene sedimentary formations (sands with clay beds), the most representative geological formation of Palmela DO. In the Óbidos DO (Centre of Portugal), two vineyards belonging to Companhia Agrícola do Sanguinhal were considered: Quinta de S. Francisco (SF; 39°11′ N, 9°10′ W) and Quinta do Sanguinhal (SA; 39°15′ N, 9°09′ W); the former is established on Eutric Regosols which developed on Lower Cretaceous sandstones, and the latter on Dystric Regosols, developed on Upper Jurassic clayey sandstones. In the Dão DO (Central Portugal), one vineyard of Sogrape Vinhos was considered: Quinta dos Carvalhais (QC; 40°33′ N, 7°47′ W), established on Dystric Cambisols, and Dystric Regosols, developed on monzonitic granites (310–350 MY), the most representative geological formation of Dão DO. The area of each vineyard, year of planting, rootstock, vine spacing, row orientation, and training system are indicated in the Table 1. All the vineyards had the same red variety in production (*Vitis vinifera* L., cv Aragonez).

The study areas have a Mediterranean-type climate. The QC, SA, and SF vineyards have a climate with dry summers and cold nights (Csb, Köppen classification), whereas the AL vineyard has a climate with dry hot summers and temperate nights (Csa, Köppen classification) [32]. Annual rainfall is variable, and in 2010, was much higher than in 2009 (for example 1598 versus 953 mm and 760 versus 1006 mm in Óbidos and Palmela DOs, respectively [33]. Soils of each vineyard show specific particle-size distribution which is associated with the respective geological formations, and their main characteristics are uniform. In the five years prior to this study, several treatment products (namely pesticides and fertilizers) were used, and may have contributed to B, Mg, Al, K, Ca, Cu, and Zn soil enrichments.

Table 1. Vineyards characteristics. (AL) Vinha de Algeruz; (SF) Quinta de S. Francisco; (SA) Quinta do Sanguinhal; (QC) Quinta dos Carvalhais.

Vineyard/ Portuguese DO	Area of the Vineyard (ha)	Year of Planting	Rootstock	Vine Spacing (m)	Row Orientation	Training System
AL vineyard/Palmela	3.0	1990	1103P	2.8 × 1.2	N-S	bi-lateral cordon
SF vineyard/Óbidos	5.0	2001	R110	2.7 × 1.0	N-S and E-W	bi-lateral cordon
SA vineyard/Óbidos	2.6	2000	R110	2.7 × 1.0	N-S	bi-lateral cordon
QC vineyard/Dão	2.5	1995	1103P	2.0 × 1.2	NE-SW	bi-lateral cordon

2.2. Soil and Grape Berries Sampling

Soil sampling took place in December 2007 (AL, SF, SA) and May 2009 (QC). Soil samples were collected with an auger from nine sampling sites distributed along three non-contiguous vine rows (representative of the entire vineyard area), considering both row and inter-row. For evaluation of soil composition heterogeneity with depth, samples were taken from five depth layers (0–20, 20–40, 40–60, 60–80 and 80–100 cm), and sealed in plastic bags (Figure 1). The samples used in the present study were those collected at the 0–20, 40–60 and 60–80 cm layers after mixing the respective row and inter-row samples.

Grape berries were sampled two weeks before the 2009 harvest. A total of 500 grape berries were collected, using plastic gloves, from the four grapevines around each soil sampling point, stored in plastic bags, and kept at cool temperatures until their processing.

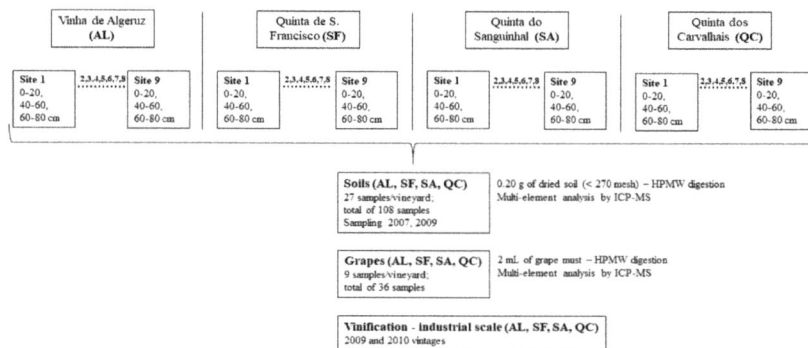

Figure 1. Schematics of the sampling strategy and analytical procedures used in this study.

2.3. Winemaking and Wine Sampling

For each vineyard, wine was prepared using all the grapes grown in the respective selected area in two consecutive harvests, 2009 and 2010. The wines were produced at three different wineries (corresponding to the wine companies participating in this study), at an industrial scale, following the traditional red winemaking process (Figure 2), with slight technological differences between wineries.

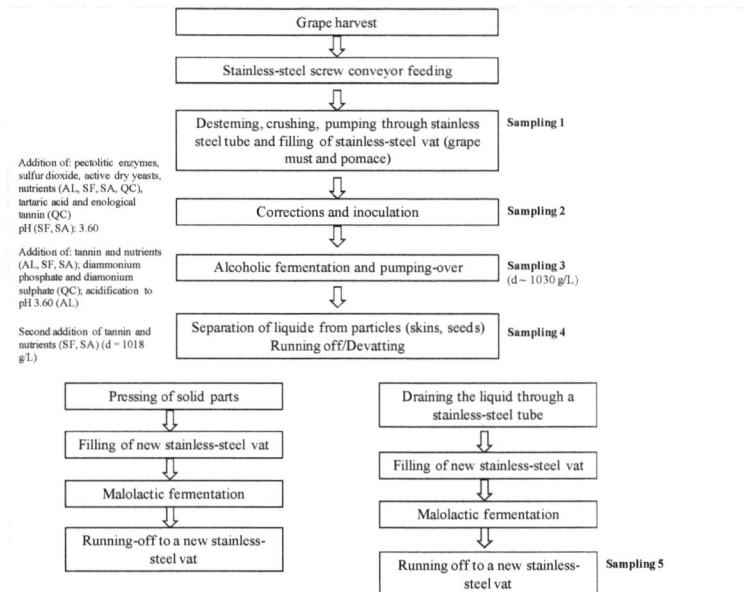

Figure 2. Schematics of the winemaking processes with indication of the points where samples were collected. (AL) Vinha de Algeruz; (SF) Quinta de S. Francisco; (SA) Quinta do Sanguinhal; (QC) Quinta dos Carvalhais.

In the 2009 vintage, samples were collected in decontaminated polyethylene tubes, in triplicate, at five different steps of the winemaking process (2009 vintage): (1) after crushing and vatting; (2) after must corrections and yeast inoculation; (3) during the alcoholic fermentation (density approximately of

1030 g/L); (4) after running off; (5) after malolactic fermentation and first racking, before any blending to preserve the trace to the vineyard of origin. Sampling 5 was repeated for the 2010 vintage.

Immediately after sampling, a volume of 5 mL of ultrapure HNO_3 was added to a sample volume of 45 mL, in order to stop the alcoholic fermentation and to start organic matter decomposition.

2.4. Sample Processing

For ICP-MS analysis, soils, grape musts and wines were treated by high pressure microwave digestion (HPMW). Soil samples were dried at room temperature, ground, and forced to pass through a sieve of 2 mm. A subsample of 0.20 g of dried soil (fraction < 0.053 mm) was acid digested using a microwave system as described by Martins et al. [9]. A certified reference material (Geo PT 25, Basalt HTB-1, International Association of Geoanalysts) was used as a reference for quality control of the analytical results.

Grape must was prepared in the laboratory by smashing the berries in plastic cups and then transferring the juice into polyethylene flasks. Immediately thereafter, a volume of 5 mL of ultrapure HNO_3 was added to a sample volume of 100 mL. Both grape musts and wines were acid digested following a digestion program previously optimized by Catarino et al. [34]. A Milestone ETHOS Plus Microwave Labstation (Milestone, Sorisole, Italy), equipped with a Milestone HPR-1000/6m monoblock high pressure rotor and TFM Teflon vessels, was used.

To avoid contamination, all polyethylene material (volumetric flasks, micropipette tips, and autosampler vessels) was immersed at least for 24 h in 20% (v/v) HNO_3, and rinsed thoroughly with purified water before use. For decontamination solution preparation, reagent grade HNO_3 was double-distilled using an infra-red subboiling distillatory system (model BSB-939-IR, Berghof, Germany). Purified water (conductivity < 0.1 μS cm^{-1}) was produced using a Seralpur Pro 90CN apparatus (Seral, Ransbach-Baumbach, Germany).

2.5. Multi-Elemental Analysis

Multi-elemental analysis was carried out with anElan 9000 ICP-MS (Perkin-Elmer SCIEX, Norwalk, CT, USA) equipped with a cross-flow nebulizer, a Ryton Scott-type spray chamber, and nickel cones. A four-channel peristaltic sample delivery pump (Gilson model) and a Perkin-Elmer AS-93 Plus autosampler (Perkin-Elmer SCIEX, Norwalk, CT, USA) protected by a laminar-flow-chamber clean room class 100 (Max Petek Reinraumtechnik, Radolfzell am Bodensee, Germany) were used. The ICP-MS instrument was controlled by Elan 6100 Windows NT software (Version 2.4, Perkin-Elmer SCIEX, Norwalk, CT, USA). The operating conditions of the ICP-MS equipment were as follows: radio-frequency (RF) power of 1200 W; Ar gas flow rates of 15 L/min for cooling, between 0.94 and 0.98 L/min for nebulizer and 1.5 L/min for auxiliary; and solution uptake rate of 1.0 mL/min.

Element concentrations were determined in mineralized soil samples (after 100-fold dilution) and in mineralized musts and wines (after 10-fold dilution), in duplicate, by adapting the ICP-MS semi-quantitative method previously described by Catarino et al. [35]. A full mass spectrum ($m/z = 6$–240, omitting the mass ranges 16–18; 40, 41, 211–229) was obtained by full mass range scanning. Rh and Re (10 μg/L) were used as internal standards.

The reference response table (Perkin-Elmer TotalQuant III, Perkin-Elmer SCIEX, Norwalk, CT, USA) was updated with different multi-elemental standard solutions with appropriate concentrations for soil analyses, and for must and wine analyses. A certified reference material (GeoPT 25, Basalt HTB-1) was periodically analysed for quality control. Also, for quality control purposes, wines from an intercomparison OIV (International Organisation of Vine and Wine) trial were analysed periodically. Between determinations, the equipment sampling system was rinsed with a 2% HNO_3 (v/v) for 75 s.

2.6. Statistical Analysis

Statistical analysis of the multi-element data was carried out to evaluate: (1) the effect of the vineyard of origin (Vineyard) on soil mineral composition; (2) the effect of soil depth level on soil mineral composition; (3) the effect of Vineyard on mineral composition of grape musts (prepared at laboratory from the grape berries sampled in the vineyards); (4) the effect of winemaking stage (Vinification) on mineral composition of wines; (5) the effect of the Vineyard on the mineral composition of wines. The statistical analysis was firstly performed by one-way analysis of variance and comparison of means (Fisher LSD, 95% level) using Statistica 7.0 software (StatSoft Inc., Tulsa, OK, USA). Normal distribution and homogeneity of variance were verified by Normal p-p (distribution of within-cell residuals) and Cochran C tests ($p < 0.05$), respectively. Whenever the parametric test assumptions were not verified, a non-parametric test (Kruskal-Wallis test) was applied. In these cases, comparison of means was not carried out. Multivariate statistical analysis, principal component analysis (PCA), and discriminant analysis (DA), was then performed using NTSYS-pc package software (Version 2.1q, Exeter Software, Setauket, NY, USA) [36] and Statistica 7.0 software (TIBCO Software Inc., Palo Alto, CA, USA), respectively. For each matrix type (soils, grape musts and wines), the results were submitted to aggregation analysis and PCA. Finally, DA was applied to the data, considering Li, Mn, Sr, and REE as variables, and the vineyards as groups.

3. Results and Discussion

3.1. Mineral Composition of Vineyard Soils

Results of the multi-element analysis and variance analysis (one-way ANOVA) of the soils of the different vineyards are listed in Table 2. It is worth mentioning that the total concentrations of the elements observed in the soils of the four vineyards are compatible with the contents of uncontaminated soils [37]. A significant effect of Vineyard on soil composition was observed for all the studied elements. As a general trend, the soil from QC (Dão DO), developed on granites, showed higher concentrations than soils from the other vineyards developed on geological sedimentary formations. This trend was especially noticeable for Li, Be, Al, Fe, Ga, Ge, As, Rb, Sn, Cs, and REE. The soil from the vineyard of Palmela DO, developed on Pliocenic sandy materials, showed the lowest concentrations for most of the elements, with Na being an exception. Data for some elements (e.g., Li, Be, Rb, Cs) in the studied vineyards are different from those reported by Almeida and Vasconcelos [4] for two vineyards from DO Douro region, installed in soils developed on schists.

Despite differences in age of the geological formations, the two vineyards from Óbidos DO showed similar concentrations for some elements, which may be explained by similar characteristics of their soils (Table S1).

Different patterns were observed in the studied vineyards (Table S2) concerning the effect of soil depth on element concentration. In fact, in soils from Palmela DO, for most of the elements, a decrease in concentration was observed with increasing depth. In contrast, for QC soil (Dão DO), no significant effect of soil depth was observed. Such a difference may be associated with the degree of soil disturbance for vineyard installation, with stronger and deeper soil disturbance occurring when a bulldozer has homogenized the different soil layers up to 100 cm depth.

Content patterns observed for Li, Be, Mn, As, Rb, Sr, and Cs suggest that these elements might be potential discriminant elements, allowing soil differentiation between the studied DOs. Indeed, similar concentrations for these elements were observed in the soils from Óbidos DO vineyards, while different concentrations were found in those from other DOs, in particular in the soil from Dão DO.

It should be emphasized that only a small proportion of total metal concentration is potentially extractable by plants, as elements are mostly strongly bonded within mineral structures. In addition, the ecotoxicology and mobility of metals in the soil depend strongly on specific chemical forms in which they are present [38].

Table 2. Multi-elemental composition (µg/g) of soils of the different vineyards. Average concentration (mean ± standard deviation). (AL) Vinha de Algeruz; (SF) Quinta de S. Francisco; (SA) Quinta do Sanguinhal; (CQ) Quinta dos Carvalhais.

Element	Vineyard Effect	AL Vineyard /Palmela	SF Vineyard /Óbidos	SA Vineyard /Óbidos	QC Vineyard /Dão
Li	**	23 ± 6	33 ± 7	39 ± 7	224 ± 50
Be	**	1.1 ± 0.3	2 ± 1	2.0 ± 0.4	31 ± 10
Na	**	4043 ± 1134	1555 ± 317	2561 ± 471	2570 ± 1081
Mg	**	737 ± 214	3226 ± 1839	1612 ± 430	5690 ± 898
Al	**	41,987 ± 9113	76,330 ± 19,270	49,982 ± 6832	149,501 ± 14,233
Ca	**	1759 ± 427	1775 ± 970	1046 ± 513	1850 ± 1126
Ti	**	5 ± 2	10 ± 3	8 ± 2	9 ± 1
V	**	20 ± 6	75 ± 29	40 ± 9	69 ± 12
Cr	**	17 ± 4	63 ± 27	29 ± 7	52 ± 12
Mn	**	87 ± 24	1147 ± 662	1166 ± 477	514 ± 105
Fe	**	6530 ± 2060	27,390 ± 11,737	12,580 ± 2877	40,450 ± 5543
Co	**	2 ± 1	15 ± 7	8 ± 2	9 ± 2
Cu	*	82 ± 39	87 ± 29	117 ± 61	79 ± 30
Ga	**	9 ± 2	19 ± 5	13 ± 2	38 ± 4
Ge	**	0.26 ± 0.06	0.31 ± 0.05	0.29 ± 0.05	0.52 ± 0.07
As	**	6 ± 3	11 ± 4	9 ± 3	46 ± 12
Rb	**	165 ± 43	190 ± 23	188 ± 27	333 ± 52
Sr	**	59 ± 15	57 ± 6	57 ± 17	34 ± 4
Sn	**	7 ± 3	6 ± 1	6 ± 2	25 ± 3
Sb	**	1.4 ± 0.9	0.8 ± 0.2	0.8 ± 0.3	0.38 ± 0.07
I	**	0.2 ± 0.1	0.6 ± 0.4	0.5 ± 0.4	0.5 ± 0.2
Cs	**	5 ± 1	9 ± 2	8 ± 1	55 ± 10
Ba	**	663 ± 316	591 ± 375	501 ± 57	305 ± 41
La	*	19 ± 5	22 ± 8	16 ± 7	23 ± 11
Ce	**	44 ± 11	49 ± 19	38 ± 15	59 ± 26
Pr	*	5 ± 1	6 ± 2	5 ± 2	7 ± 3
Nd	**	20 ± 4	22 ± 8	19 ± 7	28 ± 13
Sm	**	3.7 ± 0.9	5 ± 1	4 ± 1	6 ± 2
Eu	**	0.6 ± 0.1	1.0 ± 0.3	0.7 ± 0.1	1.0 ± 0.3
Gd	**	3 ± 1	4 ± 1	4 ± 1	6 ± 2
Tb	**	0.36 ± 0.08	0.6 ± 0.2	0.5 ± 0.1	0.9 ± 0.3
Dy	**	1.8 ± 0.4	3 ± 1	2.6 ± 0.6	5 ± 2
Ho	**	0.31 ± 0.08	0.6 ± 0.2	0.5 ± 0.1	0.9 ± 0.3
Er	**	0.9 ± 0.2	1.9 ± 0.6	1.4 ± 0.3	2.5 ± 0.8
Tm	**	0.13 ± 0.04	0.27 ± 0.08	0.21 ± 0.04	0.4 ± 0.1
Yb	**	1.0 ± 0.3	1.8 ± 0.6	1.5 ± 0.3	2.4 ± 0.6
Lu	**	0.14 ± 0.04	0.27 ± 0.08	0.23 ± 0.04	0.35 ± 0.08
W	**	0.004 ± 0.001	0.007 ± 0.002	0.006 ± 0.001	0.045 ± 0.015
Tl	**	0.9 ± 0.3	1.1 ± 0.1	1.3 ± 0.2	2.4 ± 0.3

* Significant effect ($p < 0.05$); ** significant effect ($p < 0.01$). For each vineyard and element, the results are based on average values of nine sampling sites and three depth levels (0–20 cm, 40–60 cm and 60–80 cm), in a total of 27 samples, analysed in duplicate.

For those elements significantly that were affected by Vineyard, a PCA was performed (Figure 3). The first two principal components explained 85% of the total variance. The variables which most influence the first component (C1, 70%) were Li, Be, Mg, Al, Fe, Ga, Ge, Rb, Cs, and several REE, and the variables which most influence the second component (C2, 15%) were Na and Mn. Using the first two principal components, the soils seems to separate according to the DO. The QC soils samples are well separated from the other, as well as the AL soils. Regarding the SA and SF soils, they presented a strong variability.

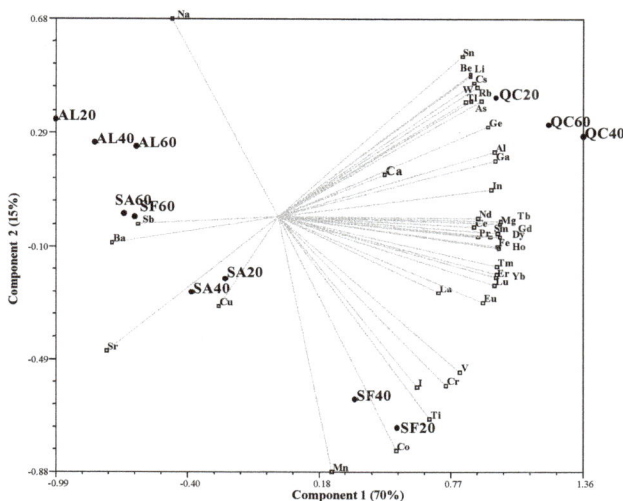

Figure 3. Principal Component Analysis performed on soils from the vineyards/DO: (AL) Vinha de Algeruz/Palmela, (SF) Quinta de S. Francisco/Óbidos, (SA) Quinta do Sanguinhal/Óbidos, and (QC) Quinta dos Carvalhais/Dão, collected at nine sampling sites and three depth levels (0–20 cm, 40–60 cm, and 60–80 cm), in a total of 27 samples, analysed in duplicate The soils are represented in the plane of the two first components which express, respectively, 70% and 15% of the total variance.

3.2. Mineral Elements Concentration during Vinification Processes

A Vineyard effect was also observed on grape must composition (data not shown, see Table S3). In fact, for most of the elements, significant differences were found between musts from different vineyards. As observed for soils, the grape must from QC showed the highest concentrations of Rb and Cs. This trend is consistent with the relationship between soil and must composition, and agrees with results reported for vineyards of Douro DO, installed in soils developed on schists [4]. However, an inverse position was observed between Óbidos and Palmela DOs, as must from AL vineyard contained higher element concentrations than those from Óbidos DO, which does not follow the trend found for the respective soils.

Several factors can modify the multi-elemental composition of grapes and wines, such as winemaking practices [10,11,13,29,30,39,40], which influence the extraction extent of mineral elements from the different parts of grape berry, element depletion by precipitation and co-precipitation phenomena, and incorporations by enological additives, among other factors. The evolution of the concentrations of elements during vinification (2009 vintage) is displayed for each vineyard (Figures 4–6), considering the following sampling steps: (1) after crushing and vatting; (2) after must corrections and inoculation; (3) during alcoholic fermentation; (4) after running off; and (5) after malolactic fermentation and first racking.

3.2.1. Alkaline and Alkaline Earth Elements

Regarding alkaline and alkaline earth elements (Figure 4), namely Li, Na, Mg, Ca, Rb, Sr, and Cs, a significant effect of Vinification was observed for all the vineyards with the exceptions of Mg, Rb, Cs, and Ba (QC). Different trends and distinct magnitudes of change were observed, depending on the element, its initial concentration, products applied and moment of application, and most probably on must-wine characteristics (pH and alcoholic strength).

With respect to Li, significant decreases occurred after corrections of the grape musts from all studied vineyards, with the greatest change (about 60%) occurring in the must/wine from the AL vineyard. These results do not agree with those from a study involving red wines, obtained by maceration of solid parts of berries where Li enrichments were likewise observed [40]. Nevertheless, the trend loss of Li during winemaking was reported by Gómez et al. [39] for white wines.

Different trends were observed between vineyards regarding Na, but without technological relevance. The increase of Na that was verified after must corrections might be explained by the addition of sodium metabisulfite.

Despite the significant effect of Vinification on Mg concentration, the small differences observed suggest its stability over time, as Mg salts are soluble, following trends reported by [39].

Concentrations of Ca tended to decrease during vinification (with exception of SF vineyard), which is associated with the low solubility of calcium tartrate, as a natural phenomenon of physical-chemical stability. The decrease (from 99 to 72 mg/L) was more evident in the must/wine from the AL vineyard, where the highest initial Ca concentration was found. The correction of pH to 3.60 with tartaric acid, performed when alcoholic fermentation was running, might be affected by Ca concentration, as the insolubility of tartaric acid salts (potassium hydrogen tartrate and calcium tartrate) is enhanced by the presence of ethanol [2].

Concentrations of Rb slightly increased during the initial phases of vinification for SF, SA, and QC musts, probably due to extraction from solid parts of the berries, favoured by maceration during alcoholic fermentation [39]. Enrichments of Sr were noticed during alcoholic fermentation as a result of its extraction from solid parts of grape berries, mostly from seeds, for musts of the SF, SA, and QC vineyards, agreeing with observations of Catarino et al. [40]. A steady concentration was observed for Cs throughout the vinification period, while Ba and especially Sr showed increasing concentrations (with the exception of AL vineyard).

Our results indicate that for most of the elements, the evolution from must to wine regarding the AL vineyard diverged from the other grape musts. In fact, strong losses were observed for Na, Mg, Rb, Sr, and Ba during the first half of alcoholic fermentation. Possibly, the increasing concentration of ethanol promotes the precipitation or co-precipitation of these metals in wine [10]. During the second phase of the alcoholic fermentation, after acid addition for pH correction, enrichments were perceived, possibly due to metal extraction from solid parts and diffusion to the must, which are favoured by the pH decrease and by the increase of ethanol content.

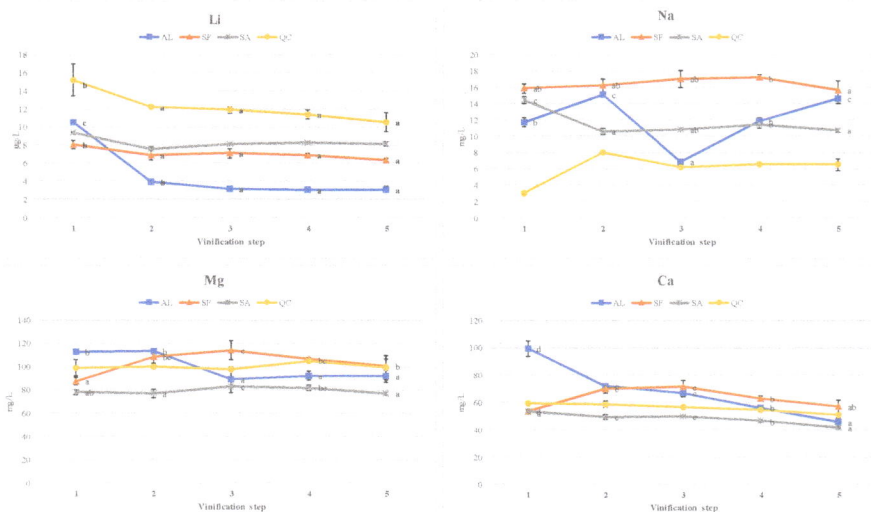

Figure 4. *Cont.*

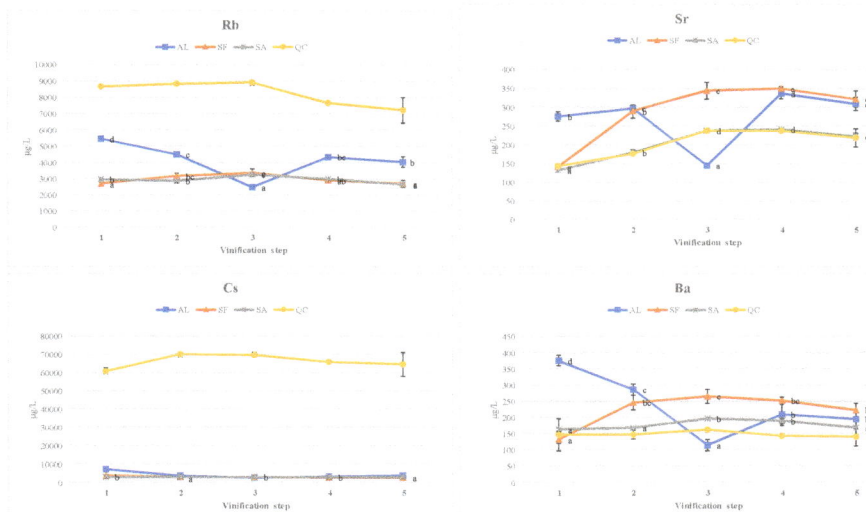

Figure 4. Evolution of alkaline and alkaline earth metals concentrations over vinification, by vineyard of origin. (AL) Vinha de Algeruz; (SF) Quinta de S. Francisco; (SA) Quinta do Sanguinhal; (QC) Quinta dos Carvalhais. Vinification step: (1) after crushing and vatting; (2) after must corrections and inoculation; (3) during the alcoholic fermentation (density approximately of 1030 g/L); (4) after running off; (5) after malolactic fermentation and first racking, before any blending to preserve the trace to the vineyard of origin. For each element and vineyard, results correspond to mean values (and corresponding standard deviations) of three replicates and corresponding analytical duplicates (*n* = 6). Means followed by the same letter are not significantly different at 0.05 level of significance. Whenever values are not followed by significance letters it means that a non-parametric parametric test (Kruskal-Wallis test) was applied and comparison of means was not carried out.

Element concentrations throughout vinification suggest the importance of their distribution in grape berries, which may be strongly influenced by technology promoting the maceration of skins and seeds. In a study focusing on the Chardonnay variety, it was observed that Ca, Sr, and Ba elements accumulate mainly in the seeds, while Li, Mg, Na, Rb, and Cs accumulate mainly in the flesh [41]. Analysing our results, the late enrichments in Sr and Ba seem consistent with kinetics extraction from seeds.

Alkaline and alkaline earth metals, namely Li, Rb, and Cs, and to a low extent Mg, Sr, and Ba, are promisor indicators of geographic origin, because they can easily be absorbed by plants in the soil. Although Mg should be considered with caution as a fingerprint, because it can be associated with correctives and soil fertilizers, it should be emphasized that wines produced in southern Portuguese soils that are rich in Mg and Na [42] showed higher concentrations of these elements as compared to other wines [21]. The higher concentrations of Li and especially of Rb and Cs in must/wine from the Dão DO compared with the other DOs suggest that these elements could be promisor fingerprints for separate Portuguese DOs.

3.2.2. Contaminant Elements

Concentrations of contaminant metals, specifically Be, Al, Mn, Fe, Ni, Cu, Zn, Ga, Mo, Sn, Sb and Tl, most of them heavy metals, are shown in Figure 5. A significant effect of Vinification was observed for all these elements and vineyards, with Zn (SF and QC), Be, Mn, and Ni (QC) being exceptions. With the exception of Mn, the concentration of contaminant elements strongly decreased during alcoholic fermentation, favoured by sulfur dioxide addition to the must, in accordance with the precipitation phenomena of heavy metals as insoluble salts, namely as sulfides, over time, as previously reported in [4,5,40].

The grape must from the vineyard AL showed for several contaminant metals, namely Al, Fe, Ni, Zn, Ga, Mo, Sb, and Tl, at the highest concentrations at step 1 of vinification, which is consistent with the results for the musts prepared at laboratory. However, no direct correspondence between must contents and final wine contents was observed, as impressive losses occurred after sulfur dioxide addition and until the moment of acidification, corroborating the important role played by sulfur dioxide as a purifying agent. In this must, depletions from step 1 to step 3 were as follows: Be-78%, Al-90%; Fe-82%; Ni-61%; Zn-75%; Ga-69%; Mo-47%; Sn-61%; Sb-78%, and Tl-71%. The depletion of these metals is positive from the wine quality perspective, given their potential participation in physical-chemical instability phenomena and potential toxicity, as excessive concentrations of Fe and Cu can cause turbidity known as ferric casse and cupric casse, respectively [2].

In contrast, slight enrichments of Be occurred during the second phase of vinification in all the musts. Surprising increases in Mn concentrations were verified in both musts from Óbidos DO after the initial corrections, suggesting contamination, which is not in agreement with a study where important losses of this metal from must to wine (red winemaking) were reported [40]. A final increase of Zn concentration in the must from AL is noteworthy; this may be of exogenous origin. In these circumstances, it is clear that contaminant elements should not be included in the group of potential markers for geographical origin fingerprints.

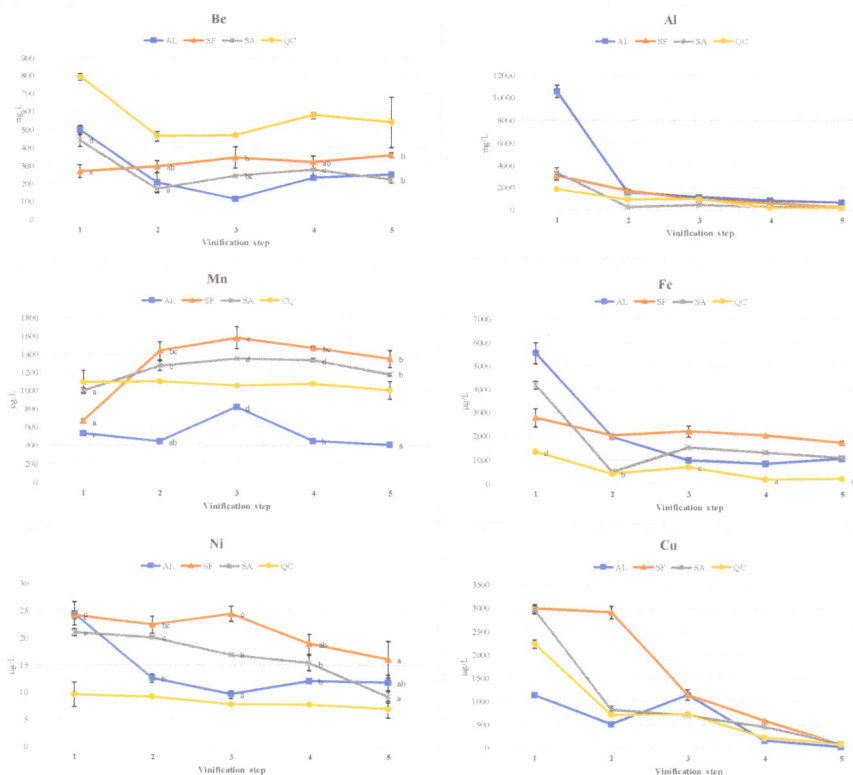

Figure 5. *Cont.*

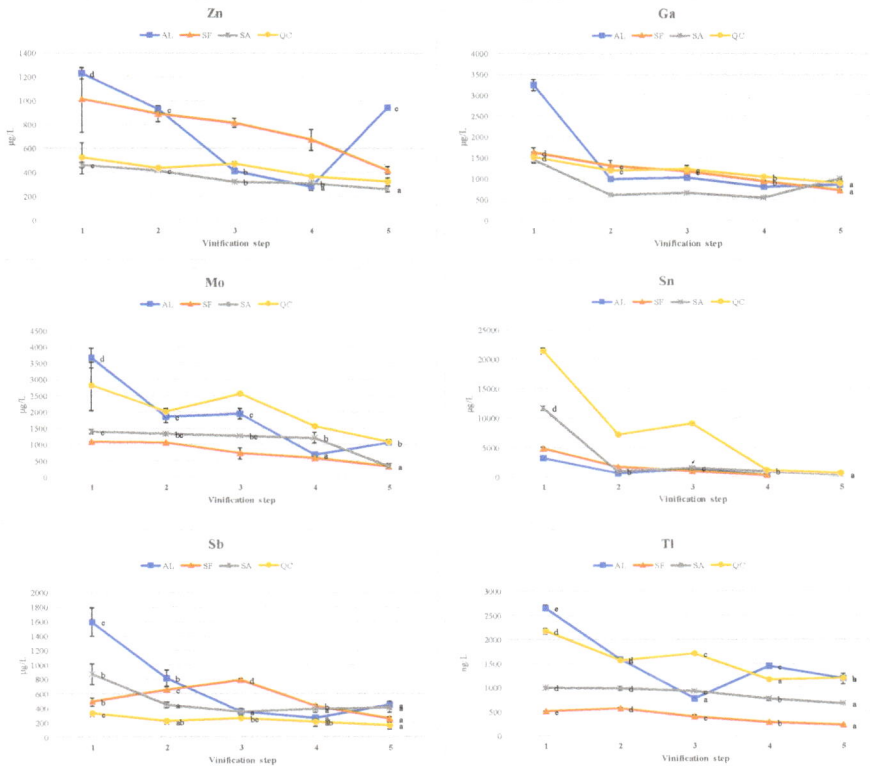

Figure 5. Evolution of heavy metals (contaminant elements) concentrations over vinification, by vineyard of origin. (AL) Vinha de Algeruz; (SF) Quinta de S. Francisco; (SA) Quinta do Sanguinhal; (QC) Quinta dos Carvalhais. Vinification step: (1) after crushing and vatting; (2) after must corrections and inoculation; (3) during the alcoholic fermentation (density approximately of 1030 g/L); (4) after running off; (5) after malolactic fermentation and first racking, before any blending to preserve the trace to the vineyard of origin. For each element and vineyard, results correspond to mean values (and corresponding standard deviations) of three replicates and corresponding analytical duplicates (*n* = 6). Means followed by the same letter are not significantly different at 0.05 level of significance. Values are not followed by significance letters indicate that a non-parametric parametric test (Kruskal-Wallis test) was applied and comparison of means was not carried out.

3.2.3. Rare Earth Elements

REE present particular interest for wine fingerprinting because, owing to their chemical similarity, the problem of selective changes of their concentration distribution is avoided. It seems that plants generally absorb REE from soil without any selectivity [43]. Furthermore, due to their chemical similarity, all the REE are expected to be affected to the same extent by insolubility and precipitation phenomena.

The present study allowed the understanding of the evolution of rare earth elements along several steps of vinification. Significant effect of Vinification on REE concentrations of the musts from the four vineyards was verified (Figure 6). As expected, for each must, a common trend was observed for all REE. In musts from the vineyards SF, SA, and QC, slight decreases occurred during vinification. In some cases, negligible increases during alcoholic fermentation could be related to REE preferential accumulation in the skin of berry [41].

The very high concentrations of REE in the must from the vineyard AL are notable in comparison with the others. In fact, the REE concentrations observed in the grape must, immediately after crushing and vatting, are in accordance with the concentrations found in grape berries (data not shown, see Table S3). Impressive decreases from step 1 to step 2 of vinification, in general higher than 70%, are most probably due to precipitation phenomena, favoured by high pH value of the musts.

It should be pointed out that no close relationship was observed between mineral content of soils and mineral content of grape berries and must at the beginning of vinification. In general, the highest concentrations were measured in the must from vineyard AL, despite the low concentrations in the corresponding soil. These results suggest that absorption of mineral elements in vineyard AL was easier than in the other vineyards, which may be facilitated by the low contents of clay and iron and aluminium compounds in the respective soil leading to lower interaction with the soil solution and facilitate element absorption.

Although the REE have been used mostly combined with other trace elements for wine fingerprinting, an excellent way to compare the REEs concentration ratios for soils, musts, and wines is by plotting REE concentrations on a chondrite-normalised diagram [6,44].

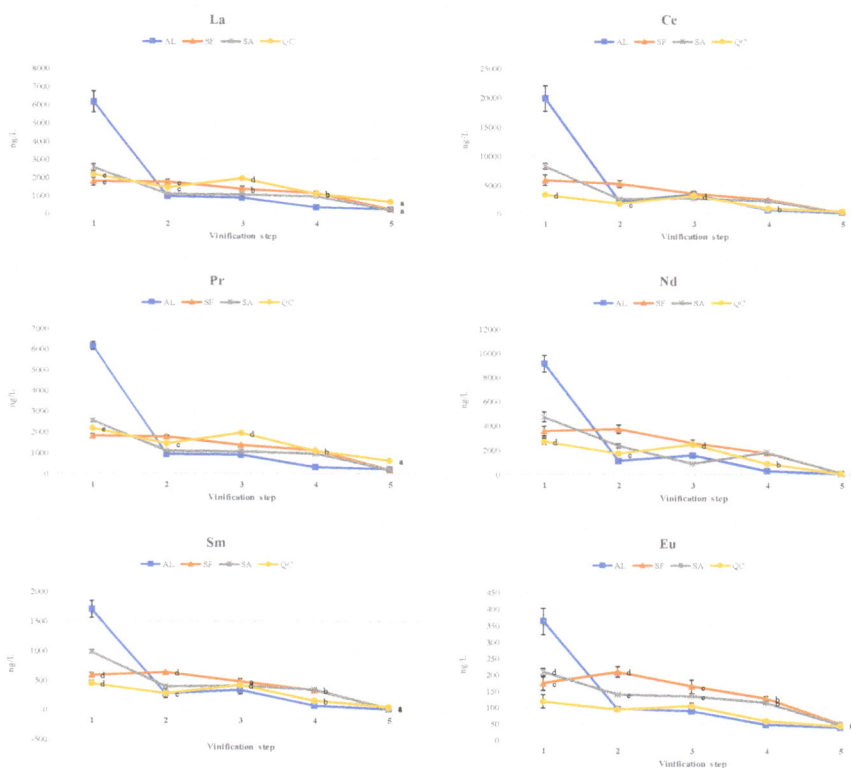

Figure 6. *Cont.*

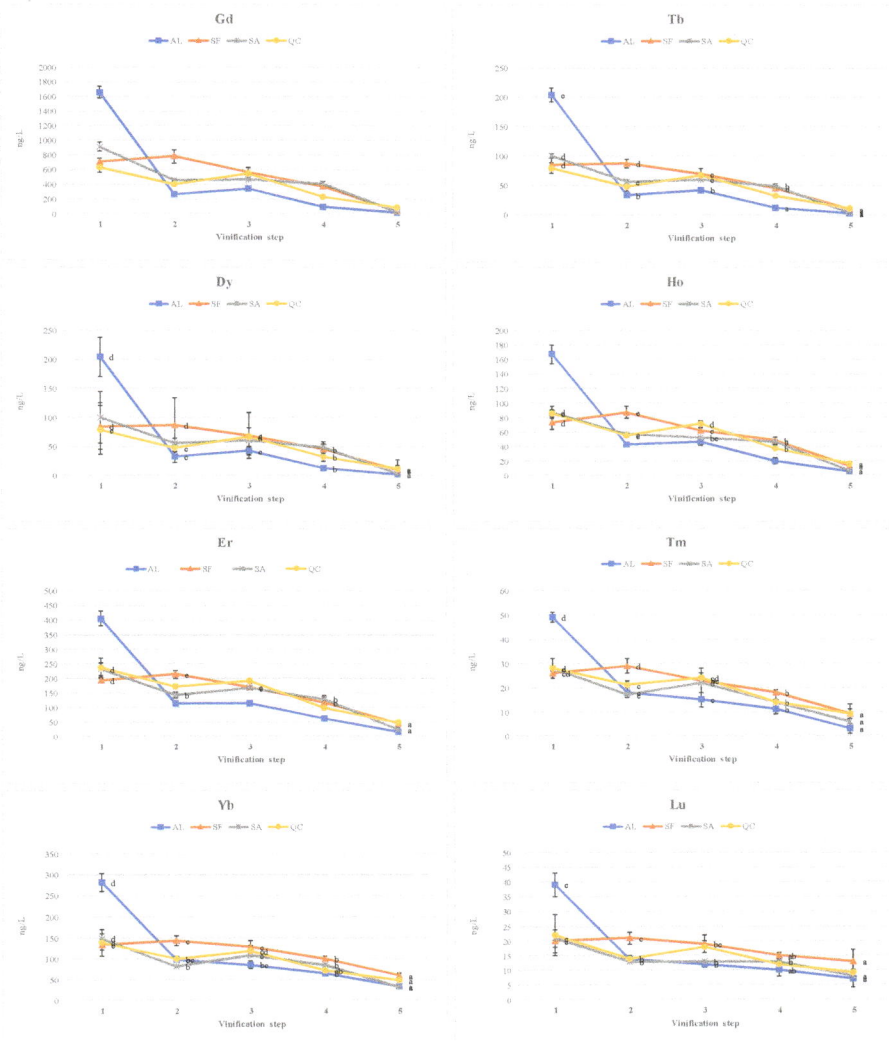

Figure 6. Evolution of rare earth elements concentrations over vinification, by vineyard of origin. (AL) Vinha de Algeruz; (SF) Quinta de S. Francisco; (SA) Quinta do Sanguinhal; (QC) Quinta dos Carvalhais. Vinification step: (1) after crushing and vatting; (2) after must corrections and inoculation; (3) during the alcoholic fermentation (density approximately of 1030 g/L); (4) after running off; (5) after malolactic fermentation and first racking, before any blending to preserve the trace to the vineyard of origin. For each element and vineyard, results correspond to mean values (and corresponding standard deviations) of three replicates and corresponding analytical duplicates (*n* = 6). Means followed by the same letter are not significantly different at 0.05 level of significance. Values not followed by significance letters indicate that a non-parametric parametric test (Kruskal-Wallis test) was applied and comparison of means was not carried out.

3.3. Mineral Composition of Wines

The mineral composition of wines is of relevant interest due to its influence on physical-chemical stability, sensory characteristics, wine safety, legal limits, and as a discriminating tool for classification [2,10].

For most of the elements, significant differences were observed between wines from different DOs (Table 3). The wine from Dão DO showed significantly higher concentrations of Li, Be, Rb, Sn, Cs, and REE than the wines from the other DOs, which is consistent with elemental contents of soils.

Table 3. Multi-elemental composition of wines (2009 vintage) from the different vineyards. (AL) Vinha de Algeruz; (SF) Quinta de S. Francisco; (SA) Quinta do Sanguinhal; (CQ) Quinta dos Carvalhais.

Element	Vineyard Effect	AL Vineyard/Palmela	SF Vineyard/Óbidos	SA Vineyard/Óbidos	QC Vineyard/Dão
Li	**	3049 ± 345 a	6323 ± 241 b	8084 ± 230 c	10,550 ± 1017 d
Be	*	251 ± 7	358 ± 17	222 ± 18	539 ± 139
Na	**	14,633 ± 639 c	15,640 ± 1166 c	10,706 ± 200 b	6480 ± 724 a
Mg	**	91,990 ± 5101 b	100,454 ± 6075 b	76,962 ± 1268 a	99,038 ± 10,380 b
Al	**	597 ± 51 c	198 ± 25 b	112 ± 6 a	129 ± 18 a
Ca	**	45,368 ± 3069 ab	56,819 ± 4786 c	41,245 ± 472 a	50,980 ± 4935 bc
Cr	*	24,997 ± 3286	18,189 ± 120	114,734 ± 1530	20,166 ± 2344
Mn	**	407 ± 21 a	1356 ± 94 d	1184 ± 14 c	1008 ± 98 b
Fe	*	1043 ± 65	1689 ± 111	1095 ± 8	186 ± 30
Ni	**	11,841 ± 1442 b	16,100 ± 3393 c	9177 ± 1085 ab	6838 ± 1563 a
Cu	*	16 ± 1	78 ± 3	92 ± 1	78 ± 16
Zn	**	940 ± 17 d	414 ± 33 c	259 ± 24 b	320 ± 32 a
Ga	*	841 ± 9	704 ± 16	984 ± 14	870 ± 100
Rb	*	4037 ± 320	2694 ± 209	2638 ± 46	7228 ± 761
Sr	**	309 ± 18 b	322 ± 22 b	222 ± 9 a	219 ± 24 a
Mo	**	1039 ± 75 b	317 ± 45 a	329 ± 80 a	1073 ± 73 b
Sn	*	137 ± 39	205 ± 27	934 ± 336	1032 ± 43
Sb	**	442 ± 54 b	250 ± 32 a	410 ± 74 b	161 ± 48 a
Cs	*	3543 ± 294	2344 ± 158	2686 ± 15	64,716 ± 6609
Ba	**	194 ± 23 bc	222 ± 21 c	167 ± 4 ab	138 ± 26 a
La	*	209 ± 79	193 ± 18	186 ± 9	634 ± 84
Ce	*	177 ± 144 a	207 ± 20 ab	123 ± 26 a	347 ± 73 b
Pr	**	20 ± 14 a	40 ± 8 b	31 ± 6 ab	85 ± 10 c
Nd	*	58 ± 57	161 ± 5	111 ± 9	347 ± 65
Sm	**	7 ± 9 a	26 ± 4 b	12 ± 6 a	46 ± 6 c
Eu	*	40 ± 6 a	50 ± 2 b	49 ± 2 ab	44 ± 8 ab
Gd	**	16 ± 12 a	34 ± 5 a	23 ± 3 a	85 ± 14 b
Tb	**	2 ± 3 a	9 ± 3 b	3 ± 1 a	11 ± 1 b
Dy	**	17 ± 15 a	43 ± 3 b	20 ± 3 a	65 ± 16 b
Ho	**	5 ± 2 a	13 ± 1 b	6 ± 2 a	16 ± 0 c
Er	**	16 ± 4 a	44 ± 7 b	23 ± 6 a	47 ± 2 b
Tm	*	3 ± 2 a	9 ± 4 b	6 ± 1 ab	9 ± 2 b
Yb	**	33 ± 4 a	59 ± 5 c	31 ± 3 a	47 ± 7 b
Lu	*	7 ± 3 a	13 ± 4 b	8 ± 1 a	9 ± 1 a
W	*	2 ± 1 ab	0.3 ± 0.2 a	1 ± 1 a	3 ± 2 b
Tl	*	1174 ± 56	206 ± 14	659 ± 6	1173 ± 109

* Significant effect ($p < 0.05$); ** significant effect ($p < 0.01$). For each vineyard and element, the average of the three replicates (considering the final winemaking step: after alcoholic fermentation and first racking) was calculated. The results are expressed as ng/L, with exception of Na, Mg, Al, Ca, Mn, Fe, Cu, Zn, Rb, Sr, and Ba, expressed as µg/L. Means followed by the same letter are not different at $p < 0.05$.

Wines from Óbidos vineyards showed similar concentrations of Rb and Cs, reflecting the soil composition. Despite the apparent higher transference of mineral elements from soil to grapes and must from Palmela vineyard, the resulting wine tends to present lower concentrations in comparison with the wines from the other vineyards. A reasonable explanation for these results can be the more intense precipitation through time favoured by the wine physical-chemical composition.

In comparison with other Portuguese DOs, wine Li concentrations in the study DOs were lower than those reported for the Douro DO (about 30 µg/L), where soils are developed on schists [4], while Mg and Mn concentrations were lower than those shown by Alentejo DO wines (close to 140 mg/L and 3 mg/L, respectively; [21]) from vineyards grown in soils rich in Mg [41]. Also, the concentrations of Li, Mg, Ca, Mn, Rb, Sr, and Cs were much higher than those reported for wines reflecting the complex of soil types in the Stellenbosch wine region [8]: 0.12–0.24 µg/L (Li), 9711–14,024 µg/L (Mg), 3789–10,343 µg/L (Ca), 94–228 µg/L (Mn), 199–630 µg/L (Rb), 41–102 µg/L

(Sr) and 0.30–6.71 µg/L (Cs). The Be and Rb may be useful to discriminate Portuguese from Romanian wines as the concentrations reported were, respectively, much higher and much lower [22] than those measured in the present study. Moreover, Rb concentrations are higher compared with those of German wines (160 to 970 µg/L), from Baden, Rheingau, Rheinhessen, and Pfalz wine regions [5].

Concentrations of Cu and Zn were very low in all the wines, and much lower than the OIV maximum acceptable limits of 1 and 5 mg/L respectively [45]. In fact, for all elements, concentrations are in accordance with the respective normal variation range, considering the grape variety and winemaking technology [2,10].

For those elements where a significant effect of Vineyard was found, PCA was performed, the results being displayed in Figure 7. The projection of the wines concerning the two vintages (2009 and 2010) in the C1-C2 plane was consistent with the projection of soils. The first two principal components, generated from PCA analysis of the data, explained 69.5% of the total variance. The variables which influence most the first component (C1, 47.5%) were Mn and REE, and the variables which influence most the second component (C2, 22.0%) Li, Be, Rb, and Cs. The projection of the wines in the C1-C2 plane presents some similarities with the projection of soils (Figure 3).

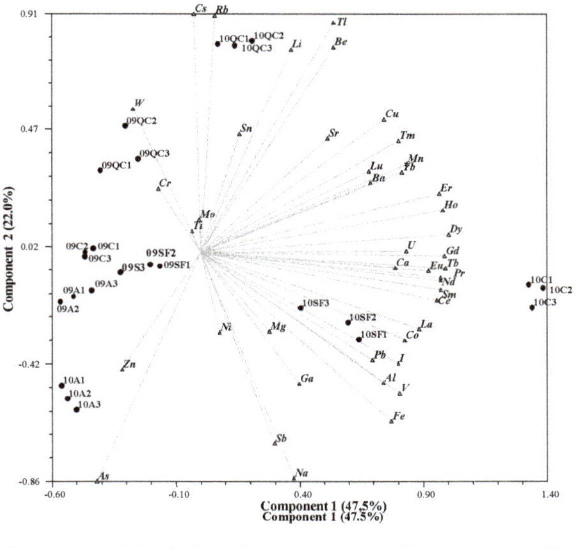

Figure 7. Principal Component Analysis performed on elemental wine composition (2009 and 2010 vintages) from the vineyards/DO: (AL) Vinha de Algeruz/Palmela, (SF) Quinta de S. Francisco/Óbidos, (SA) Quinta do Sanguinhal/Óbidos, and (QC) Quinta dos Carvalhais/Dão, collected at the final step of the vinification processes (5), in triplicate (1, 2, 3). Wines are represented in the plane of the two first components which express, respectively, 47.5% and 22.0% of the total variance.

3.4. Vineyards and Vintage Year

Results concerning the two vintages, 2009 (Table 3) and 2010 (Table S4) suggest the influence of the vintage year on wine elemental composition, as different hydrological conditions occurred during the study period; for instance, the annual rainfall in 2009 (especially during spring) was lower than the climatological normal, whereas in 2010 was much higher. Yet, with regards to DA results, considering both the wines from 2009 and 2010 vintages (six samples from each vineyard), Li, Mn, Sr, and REE allowed wine discrimination according to vineyard/geographical origin, suggesting their robustness to vintage effect (Figure 8). These results agree with findings of a study involving Australian wines from 19 vintages, which indicated that the multi-element composition is essentially independent of their vintage [7]. Also, they follow the trend observed for REE in wines from different vintages and from the same vineyard over a period of several years [44].

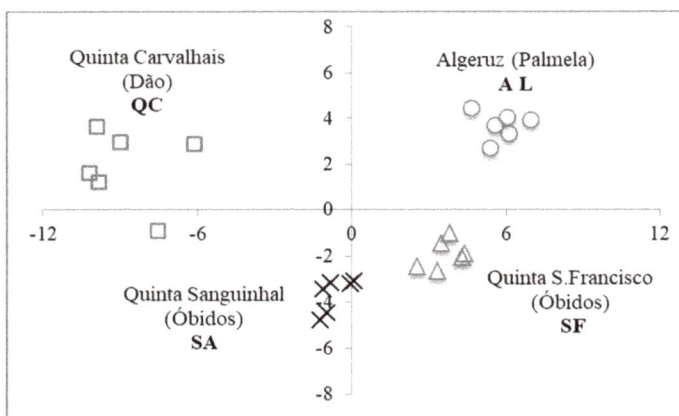

Figure 8. Scores plot of discriminant analysis (DA) of wine samples (2009 and 2010 vintages, vinification step 5) from different vineyards/DO, using Li, Mn, Rb, Sr and REE as variables.

Despite the controversy associated with their potential contamination [11,46], it is worthwhile to emphasize the role played by REE as discriminant elements [6,20]. This role can be better explored by the chondrite-normalisation approach, which allowed strong correlation to be observed between the REE patterns of grape musts and those of the provenance soils from Portuguese DOs [6]. Also, the REE patterns of wines from different vineyards in France, California, and Australia revealed inter-regional variations [44].

Despite the elemental concentration variations observed throughout vinification, the elemental composition of the wines reflected the elemental composition of the provenance soils for some elements. Differences were observed between the multi-elemental compositions of wines from different Portuguese vineyards/geographical origins, which enabled their successful discrimination through the application of multivariate statistics. Furthermore, the potential applicability of this strategy for intra-regional classification of wines was demonstrated. The results of this study obtained by application of a geochemical approach represent a valuable contribution both for viticultural zoning and for the building a database concerning Portuguese wines. Nevertheless, further research should be performed involving all the representative lithological formations of the studied DOs in order to characterize each region, as well as other Portuguese DOs. Moreover, the influence of the vintage year, in direct relationship especially with climate changes, requires further research.

Supplementary Materials: The following are available online at http://www.mdpi.com/2306-5710/4/4/85/s1, Table S1: Mean (±1 SD, *n* = 9) values of coarse fragments >2 mm (CF), sand (SD), silt (SI), clay (CL), pH, organic carbon (orgC), base cations and sum of bases (SB), phosphorus by Olsen test (P$_{OL}$), iron and aluminium determined by the dithionite-sodium citrate method in the samples from the vineyards of Algezuz (AL), Quinta de São Francisco (SF), Quinta do Sanguinhal (SA), and Quinta de Carvalhais (QC) at 0–20, 40–60 and 60–80 cm depth; Table S2: Effect of depth level on soil multi-element composition (µg/g): (a) Algeruz (AL) vineyard; (b) Quinta de S. Francisco (SF) vineyard; (c) Quinta do Sanguinhal (SA) vineyard; (d) Quinta dos Carvalhais (QC) vineyard; Table S3: Vineyard effect on multi-elemental composition of grape berries (corresponding musts) collected in the vineyards of Algezuz (AL), Quinta de São Francisco (SF), Quinta do Sanguinhal (SA), and Quinta de Carvalhais (QC); Table S4: Multi-elemental composition of wines (2010 vintage) from the different vineyards. (AL) Vinha de Algeruz; (SF) Quinta de S. Francisco; (SA) Quinta do Sanguinhal; (CQ) Quinta dos Carvalhais.

Author Contributions: S.C., A.C.-G., M.M. and R.B.d.S. conceived and designed the experiments; S.C. and F.M. performed the experiments; S.C., M.M. and I.C. analyzed the data; S.C., A.C.-G., M.M. and R.B.d.S. contributed reagents/materials/analysis tools; S.C. and M.M. wrote the original draft; S.C., M.M., I.C., A.C.-G. and R.B.d.S. revised the paper.

Funding: This research was supported by the Portuguese National Funding Agency for Science and Technology through the R&D project PTDC/AGR-ALI/64655/2006, grant SFRH/BPD/93535/2013, and through the research centre LEAF (UID/AGR/04129/2013).

Acknowledgments: The authors acknowledge the enterprises Companhia Agrícola do Sanguinhal Lda, José Maria da Fonseca Vinhos, and Sogrape Vinhos for providing their facilities regarding the project development; Paulo Marques and José Correia for help in field sampling; the staff of the Soil Laboratory (ISA, Lisbon) for soil analysis; and Otília Cerveira for help in Mineral Analysis Laboratory activities (INIAV, Dois Portos).

Conflicts of Interest: The authors declare no conflict of interest.

References

1. OIV. *Traceability Guidelines in the Vitivinicultural Sector*; Resolution OIV CST 1/2007; International Organisation of Vine and Wine: Paris, France, 2007.
2. Ribéreau-Gayon, P.; Dubourdieu, D.; Donèche, B.; Lonvaud, A. *Handbook of Enology. The Chemistry of Wine. Stabilization and Treatments*; John Wiley & Sons: Chichester, UK, 2006.
3. Greenough, J.D.; Longerich, H.P.; Jackson, S.E. Element fingerprinting of Okanagan Valley wines using ICP-MS: Relationships between wine composition, vineyard and wine colour. *Aust. J. Grape Wine Res.* **1997**, *3*, 75–83. [CrossRef]
4. Almeida, C.M.R.; Vasconcelos, M.T.S.D. Multielement composition of wines and their precursors including provenance soil and their potentialities as fingerprints of wine origin. *J. Agric. Food Chem.* **2003**, *51*, 4788–4798. [CrossRef] [PubMed]
5. Gómez, M.D.M.C.; Feldmann, I.; Jakubowski, N.; Andersson, J.T. Classification of German white wines and certified brand of origin by multielement quantitation and pattern recognition techniques. *J. Agric. Food Chem.* **2004**, *52*, 2962–2974.
6. Catarino, S.; Trancoso, I.M.; Madeira, M.; Monteiro, F.; Bruno de Sousa, R.; Curvelo-Garcia, A.S. Rare earths data for geographical origin assignment of wine: A Portuguese case study. *Bull. OIV* **2011**, *84*, 233–246.
7. Martin, A.E.; Watling, R.J.; Lee, G.S. The multi-element determination of Australian wines. *Food Chem.* **2012**, *133*, 1081–1089. [CrossRef]
8. Coetzee, P.P.; Van Jaarsveld, F.P.; Vanhaecke, F. Intraregional classification of wine via ICP-MS elemental fingerprinting. *Food Chem.* **2014**, *164*, 485–492. [CrossRef] [PubMed]
9. Martins, P.; Madeira, M.; Monteiro, F.; Bruno de Sousa, R.; Curvelo-Garcia, A.S.; Catarino, S. ^{87}Sr/^{86}Sr ratio in vineyards soils from Portuguese Designations of Origin and its potential for provenance authenticity. *J. Int. Sci. Vigne Vin.* **2014**, *48*, 21–29.
10. Catarino, S.; Curvelo-Garcia, A.S.; Bruno de Sousa, R. Contaminant elements in wines: A review. *Ciência Téc. Vitiv.* **2008**, *23*, 3–19.
11. Catarino, S.; Madeira, M.; Monteiro, F.; Rocha, F.; Curvelo-Garcia, A.S.; Bruno de Sousa, R. Effect of bentonite characteristics on the elemental composition of wine. *J. Agric. Food Chem.* **2008**, *56*, 158–165. [CrossRef] [PubMed]
12. Médina, B.; Augagneur, S.; Barbaste, M.; Grousset, F.E.; Buat-Ménard, P. Influence of atmospheric pollution on the lead content of wines. *Food Addit. Contam.* **2000**, *6*, 435–445. [CrossRef] [PubMed]
13. Nicolini, G.; Larcher, R.; Pangrazzi, P.; Bontempo, L. Changes in the contents of micro- and trace-elements in wine due to winemaking treatments. *Vitis* **2004**, *43*, 41–45.
14. Volpe, M.G.; La Cara, F.; Volpe, F.; De Mattia, A.; Serino, V.; Petitto, F.; Zavalloni, C.; Limone, F.; Pellechia, R.; De Prisco, P.P.; et al. Heavy metal uptake in the enological food chain. *Food Chem.* **2009**, *117*, 553–560. [CrossRef]
15. Fabani, M.P.; Toro, M.E.; Vázquez, F.; Díaz, M.P.; Wunderlin, D.A. Differential absorption of metals from soil to diverse vine varieties from the Valley of Tulum (Argentina): Consequences to evaluate wine provenance. *J. Agric. Food. Chem.* **2009**, *57*, 7409–7416. [CrossRef] [PubMed]
16. Wilkes, E.; Day, M.; Herderich, M.; Johnson, D. AWRI report: In vino veritas—Investigating technologies to fight wine fraud. *Wine Vitic. J.* **2016**, *31*, 36–38.
17. Dutra, S.V.; Adami, L.; Marcon, A.R.; Carnieli, G.J.; Roani, C.A.; Spinelli, F.R.; Leonardelli, S.; Ducatti, C.; Moreira, M.Z.; Vanderlinde, R. Determination of the geographical origin of Brazilian wines by isotope and mineral analysis. *Anal. Bioanal. Chem.* **2011**, *401*, 1571–1576. [CrossRef] [PubMed]

18. Greenough, J.D.; Mallory-Greenough, L.M.; Fryer, B.J. Geology and wine: Regional trace element fingerprinting of Canadian wines. *Geosci. Can.* **2005**, *32*, 129–137.

19. Galgano, F.; Favati, F.; Caruso, M.; Scarpa, T.; Palma, A. Analysis of trace elements in southern Italian wines and their classification according to provenance. *LWT Food Sci. Technol.* **2008**, *41*, 1808–1815. [CrossRef]

20. D'Antone, C.; Punturo, R.; Vaccaro, C. Rare earth elements distribution in grapevine varieties grown on volcanic soils: An example from Mount Etna (Sicily, Italy). *Environ. Monit. Assess.* **2017**, *189*, 160. [CrossRef] [PubMed]

21. Rodrigues, S.M.; Otero, M.; Alves, A.A.; Coimbra, J.; Coimbra, M.A.; Pereira, E.; Duarte, A.C. Elemental analysis for categorization of wines and authentication of their certified brand of origin. *J. Food Compos. Anal.* **2011**, *24*, 548–562. [CrossRef]

22. Geana, I.; Iordache, A.; Ionete, R.; Marinescu, A.; Ranca, A.; Culea, M. Geographical origin identification of Romanian wines by ICP-MS elemental analysis. *Food Chem.* **2013**, *138*, 1125–1134. [CrossRef] [PubMed]

23. Dinca, O.R.; Ionete, R.E.; Costinel, D.; Geana, I.E.; Popescu, R.; Stefanescu, I.; Radu, G.L. Regional and vintage discrimination of Romanian wines based on elemental and isotopic fingerprinting. *Food Anal. Methods* **2016**, *9*, 2406–2417. [CrossRef]

24. Coetzee, P.P.; Steffens, F.E.; Eiselen, R.J.; Augustyn, O.P.; Balcaen, L.; Vanhaecke, F. Multi-element analysis of South African wines by ICP-MS and their classification according to geographical origin. *J. Agric. Food Chem.* **2005**, *53*, 5060–5066. [CrossRef] [PubMed]

25. Minnaar, P.P.; Rohwer, E.R.; Booyse, M. Investigating the use of element analysis for differentiation between the geographic origins of Western Cape wines. *S. Afr. J. Enol. Vitic.* **2005**, *26*, 95–105. [CrossRef]

26. Herrero-Latorre, C.; Medina, B.J. Utilisation de quelques éléments minéraux dans la différenciation des vins de Galice de ceux d'autres régions d'Espagne. *J. Int. Sci. Vigne Vin.* **1990**, *24*, 147–156.

27. Gonzálvez, A.; Llorens, A.; Cervera, M.L.; Armenta, S.; de la Guardia, M. Elemental fingerprint of wines from the protected designation of origin Valencia. *Food Chem.* **2009**, *112*, 26–34. [CrossRef]

28. Di Paola-Naranjo, R.D.; Baroni, M.V.; Podio, N.S.; Rubinstein, H.R.; Fabani, M.P.; Badini, R.G.; Inga, M.; Ostera, H.A.; Cagnoni, M.; Gallegos, E.; et al. Fingerprints for main varieties of Argentinean wines: Terroir differentiation by inorganic, organic, and stable isotopic analysis coupled to chemometrics. *J. Agric. Food Chem.* **2011**, *59*, 7854–7865. [CrossRef] [PubMed]

29. Moreira, C.; de Pinho, M.; Curvelo-Garcia, A.S.; Bruno de Sousa, R.; Ricardo-da-Silva, J.M.; Catarino, S. Evaluating nanofiltration effect on wine ^{87}Sr/^{86}Sr isotopic ratio and the robustness of this geographical fingerprint. *S. Afr. J. Enol. Vitic.* **2017**, *38*, 82–93. [CrossRef]

30. Kaya, A.; Bruno de Sousa, R.; Curvelo-Garcia, A.S.; Ricardo-da-Silva, J.; Catarino, S. Effect of wood aging on mineral composition and wine ^{87}Sr/^{86}Sr isotopic ratio. *J. Agric. Food Chem.* **2017**, *65*, 4766–4776. [CrossRef] [PubMed]

31. OIV. *World Vitiviniculture Situation 2016, OIV Statistical Report on Wine Vitiviniculture*; International Organisation of Vine and Wine: Paris, France, 2016.

32. IPMA, Instituto Português do Mar e da Atmosfera (Portuguese Government). Available online: https://www.ipma.pt/pt/ (accessed on 7 November 2018).

33. Instituto Nacional de Estatística. Statistics Portugal. Available online: www.ine.pt (accessed on 7 November 2018).

34. Catarino, S.; Trancoso, I.M.; Bruno de Sousa, R.; Curvelo-Garcia, A.S. Grape must mineralization by high pressure microwave digestion for trace element analysis: Development of a procedure. *Ciência e Técnica Vitivinícola* **2010**, *25*, 87–93.

35. Catarino, S.; Curvelo-Garcia, A.S.; Bruno de Sousa, R. Measurements of contaminant elements of wines by inductively coupled plasma mass spectrometry: A comparison of two calibration approaches. *Talanta* **2006**, *70*, 1073–1080. [CrossRef] [PubMed]

36. Rohlf, F.J. *NTSYS-pc: Numerical Taxonomy and Multivariate Analysis System*; Exeter Software: New York, NY, USA, 2000; pp. 18–31.

37. CMCE. *Canadian Soil Quality Guidelines for the Protection of Environmental and Human Health*; Update 7.0; Canadian Council of Ministers of the Environment: Quebec, QC, Canada, 2007; pp. 1–6.

38. Brady, N.C.; Weil, R.R. *The Nature and Properties of Soils*, 14th ed.; Pearson International Edition; Pearson Education, Inc.: Upper Saddle River, NJ, USA, 2008.

39. Gómez, M.D.M.C.; Brandt, R.; Jakubowski, N.; Anderson, J.T. Changes of the metal composition in German white wines through the winemaking process. A study of 63 elements by inductively coupled plasma-mass spectrometry. *J. Agric. Food Chem.* **2004**, *52*, 2953–2961.

40. Catarino, S.; Capelo, J.L.; Curvelo-Garcia, A.S.; Bruno de Sousa, R. Evaluation of contaminant elements in Portuguese wines and original musts by inductively coupled plasma mass spectrometry. *J. Int. Sci. Vigne Vin.* **2006**, *40*, 91–100.

41. Bertoldi, D.; Larcher, R.; Bertamini, M.; Otto, S.; Concheri, G.; Nicolini, G. Accumulation and distribution pattern of macro- and microelements and trace elements in *Vitis vinifera* L. cv. Chardonnay berries. *J. Agric. Food Chem.* **2011**, *59*, 7224–7236. [CrossRef] [PubMed]

42. Monteiro, F.M.G. Factores Determinantes do Hidromorfismo em Solos do Sul de Portugal. Ph.D. Thesis, Universidade Técnica de Lisboa, Lisboa, Portugal, 2004.

43. Cao, X.; Chen, Y.; Gu, Z.; Wang, X. Determination of trace rare earth elements in plant and soil samples by inductively coupled plasma-mass spectrometry. *Int. J. Environ. Anal. Chem.* **2000**, *76*, 295–309. [CrossRef]

44. Augagneur, S.; Médina, B.; Szpunar, J.; Lobinski, R. Determination of rare earth elem1ents in wine by inductively coupled plasma mass spectrometry using a microconcentrc nebulizer. *J. Anal. At. Spectrom.* **1996**, *11*, 713–721. [CrossRef]

45. OIV. *Compendium of International Methods of Wine and Must Analysis*; International Organisation of Vine and Wine: Paris, France, 2017.

46. Jakubowski, N.; Brandt, R.; Stuewer, D.; Eschnauer, H.; Görtges, S. Analysis of wines by ICP-MS: Is the pattern of the rare earth elements a reliable fingerprint for the provenance? *Fresenius J. Anal. Chem.* **1999**, *364*, 424–428. [CrossRef]

Article

High-Resolution Mass Spectrometry Identification of Secondary Metabolites in Four Red Grape Varieties Potentially Useful as Traceability Markers of Wines

Christine M. Mayr [1,2], Mirko De Rosso [1], Antonio Dalla Vedova [1] and Riccardo Flamini [1,*]

[1] Council for Agricultural Research and Economics-Viticulture & Enology (CREA-VE), Viale XXVIII Aprile 26, 31015 Conegliano (TV), Italy; mayrchristine@gmx.de (C.M.M.); mirko.derosso@crea.gov.it (M.D.R.); tonidallavedova@tin.it (A.D.V.)

[2] Department of Agronomy, Food, Natural Resources, Animals and Environment (DAFNE), University of Padova, 35020 Legnaro (PD), Italy

* Correspondence: riccardo.flamini@crea.gov.it; Tel.: +39-0438-456749

Received: 3 August 2018; Accepted: 19 September 2018; Published: 5 October 2018

Abstract: Liquid chromatography coupled to high-resolution mass spectrometry (LC-Q/TOF) is a powerful tool to perform chemotaxonomic studies through identification of grape secondary metabolites. In the present work, the metabolomes of four autochthonous Italian red grape varieties including the chemical classes of anthocyanins, flavonols/flavanols/flavanones, and terpenol glycosides, were studied. By using this information, the metabolites that can potentially be used as chemical markers for the traceability of the corresponding wines were proposed. In Raboso wines, relatively high abundance of both anthocyanic and non-anthocyanic acyl derivatives, is expected. Potentially, Primitivo wines are characterized by high tri-substituted flavonoids, while Corvina wines are characterized by higher di-substituted compounds and lower acyl derivatives. Negro Amaro wine's volatile fraction is characterized by free monoterpenes, such as α-terpineol, linalool, geraniol, and Ho-diendiol I. A similar approach can be applied for the traceability of other high-quality wines.

Keywords: wine; grape; traceability; metabolomics; high-resolution mass spectrometry; Amarone; Recioto; Raboso; Primitivo; Negro Amaro

1. Introduction

Amarone della Valpolicella and Recioto are two red DOCG wines (controlled and guaranteed designation of origin) produced in Northeast Italy (Verona province, Veneto) by using a blend of autochthonous red grape varieties, such as Corvina Veronese and Corvinone. Types and percentages of grape varieties that can be used are stated in the disciplinary of production of the wines (approved by Ministerial Decree 24 March 2010), which defines the municipalities allowed for the cultivation, the maximum yield per hectare, and the winemaking practices allowed. The main variety used is Corvina Veronese, which has to account for 45–95% of the grape blend.

Raboso Piave is another red grape variety cultivated in the Veneto region, whose grapes are characterized by high polyphenolic content, used to produce the high-quality reinforced wine *Raboso Passito* DOCG [1].

Primitivo and Negro Amaro are two red grape varieties cultivated in Southern Italy. In general, these grapes are characterized by high sugar and polyphenolic content and the corresponding wines by high alcohol and color [2–4].

Despite the measures in place to regulate and guarantee the authenticity and geographical traceability of wines, different kinds of fraud (e.g., mislabeling, blending with wines of a lesser quality and/or without denomination of origin, etc.) has been reported [5]. In this context, over the last years a growing interest in developing analytical methods for wine authentication has been observed [6,7]. For the characterization of wine origin and variety, as well as the grape growing and winemaking practices used, the chemical characterization of wines is generally based on the characterization of the polyphenolic compounds, such as anthocyanins, flavones, flavonols, hydroxycinnamic acids, as well as of aroma compounds, such as terpenols, norisoprenoids, and benzenoids [8–14].

Among the metabolomic methods available, liquid chromatography coupled to high-resolution mass spectrometry (HRMS) is very effective by providing the identification of several hundred metabolites in grape and wine in just two analyses performed in positive and negative ionization modes [15–19]. Recently, an approach of HRMS-suspect screening metabolomics in grape was proposed and it allowed identification of new grape compounds belonging to the chemical classes of stilbenes, flavonols, anthocyanins, and glycoside terpenes [20–22].

In the present study this method was used to investigate the metabolome of Corvina, Raboso Piave, Primitivo, and Negro Amaro grapes. In particular, the profiles of flavonols, flavanols and flavanones, glycoside terpenols, procyanidins, stilbenes, and anthocyanins of each variety were determined, and the peculiar metabolites, which can be used as traceability markers of the corresponding wines, were identified.

2. Materials and Methods

2.1. Samples and Standards

Grape samples of *Vitis vinifera* Corvina Veronese, Primitivo, and Negro Amaro were harvested in 2016, while Raboso Piave grapes were collected in 2013. All samples were sourced from the vine Germoplasm Collection of the CREA-Viticulture & Enology sited in Susegana (Veneto, Italy). For each variety, 100 berries were collected at the technological maturity (maximum soluble solid content in the juice) from five different plants using randomized criteria, and kept frozen at −20 °C until analysis.

Standards of kaempferol-3-*O*-glucoside, quercetin-3-*O*-glucoside, myricetin-3-*O*-glucoside, malvidin-3-*O*-glucoside, kaempferol-3-*O*-glucuronide, (−)-epicatechin, (+)-catechin, (−)-epigallocatechin, procyanidin B1, procyanidin B2, tamarixetin, syringetin, and rutin were purchased from Extrasynthese (Genay, France); quercetin, myricetin, kaempferol, *trans*-resveratrol, *trans*-piceid, piceatannol, E-piceid, isorhamnetin, and 4′,5,7-trihydroxy flavanone from Sigma-Aldrich (Milan, Italy). δ-viniferin was provided by CT Chrom (Marly, Switzerland). Z-piceid was produced by photoisomerization of the *E* isomer as reported for the isomerization of *trans*-resveratrol (around 80% conversion yield) [23]. E-ε-viniferin was extracted from lignified vine cane as proposed by Pezet and coworkers [24].

2.2. Sample Preparation

Sample preparation for analysis was performed using 20 grape berries. After removing the seeds, pulp and skins were ground under liquid nitrogen using an ultra-turrax (IKA, Staufen, Germany). Pure methanol was added to the resulting powder in a ratio 2:1 (*v*/*w*), and the extraction was carried out for 20 min. After the addition of 200 μL of 4′,5,7-trihydroxyflavanone solution (520 mg/L) as internal standard, samples were centrifuged (2957 rcf, 18 °C, 12 min), the supernatant was filtered by using an Acrodisc GHP 0.22 μm filter (Waters, Milford, MA, USA) and LC/MS analysis of the solution was performed. For each variety (Corvina, Primitivo, Negro Amaro, and Raboso Piave), two grape samples were studied.

2.3. UHPLC-Q/TOF Analysis

An analytical system composed by Ultra-High Performance Liquid Chromatography (UHPLC) Agilent 1290 Infinity coupled to Agilent 1290 Infinity G4226A autosampler and accurate-mass Quadrupole-Time of Flight (Q/TOF) Mass Spectrometer Agilent 6540 (nominal resolution 40000) with Agilent Dual Jet Stream Ionization source (Agilent Technologies, Santa Clara, CA, USA), was used.

Data acquisition software: Agilent MassHunter version B.04.00 (B4033.2). Chromatographic separation was performed by Zorbax reverse-phase column (RRHD SB-C18 3 × 150 mm, 1.8 μm) (Agilent Technologies, Santa Clara, CA, USA) using solvent A 0.1% (*v*/*v*) aqueous formic acid and solvent B 0.1% (*v*/*v*) formic acid in acetonitrile, and the following elution gradient program: 5% B isocratic for 8 min, from 5% to 45% B in 10 min, from 45% to 65% B in 5 min, from 65% to 90% in 4 min, 90% B isocratic for 10 min; flow rate 0.4 mL/min. Sample injection 5 μL; column temperature 35 °C. False positives were checked by analyzing a blank between each pair of samples. For each sample, two repeated analyses in both positive and negative ionization mode were performed.

Q/TOF conditions: sheath gas nitrogen 10 L/min at 400 °C; drying gas nitrogen 8 L/min at 350 °C; nebulizer pressure 60 psig, nozzle voltage 0 kV (negative ionization mode) and 1 kV (positive ionization mode), capillary voltage ±3.5 kV in positive and negative ion modes, respectively. Signals in the m/z 100–1700 range, were recorded. Mass calibration was performed with standard mix G1969-85000 (Supelco Inc.) and had residual error for the expected masses between ±0.2 ppm. Lock masses: TFA anion at m/z 112.9856 and HP-0921(+formate) at m/z 966.0007 in negative-ion mode, purine at m/z 121.0509 and HP-0921 at m/z 922.0098 in positive-ion mode.

Data analysis was performed by Agilent MassHunter Qualitative Analysis software version B.05.00 (5.0.519.0). Compound identification was based on accurate mass and isotope pattern and expressed as "overall identification score" computed as weighted average of the isotopic pattern signal (W_{mass} = 100, $W_{abundance}$ = 60, $W_{spacing}$ = 50, mass expected data variation 2.0 mDa + 5.6 ppm, mass isotope abundance 7.5%, mass isotope grouping peak spacing tolerance 0.0025 m/z + 7.0 ppm).

Targeted data analysis was performed by using the algorithm '*Find by Molecular Formula*'. Compounds were identified by using the in-house constructed HRMS database *GrapeMetabolomics*. Identifications were confirmed by performing autoMS/MS of the precursor ions in the m/z 100–1700 range (collision energy 20–60 eV, acquisition rate 2 spectra/s) and using the standards available.

2.4. Statistical Analysis

Multivariate analysis was performed by using the $[M − H]^-$ or $[M]^+$ ion peak area normalized to the internal standard.

Tukey's test was performed by PAST 3.01 software (Paleontological statistics software package for education and data analysis; Hammer, Ø., Harper, D.A.T., Ryan, P.D. 2001) using the intensity of the normalized recorded signals. The data with different letters were significantly different for $p < 0.01$.

Principal component and Cluster analyses (Ward method, Euclidean distance) were performed by MetaboAnalyst, version 4.0 (http://www.metaboanalyst.ca, last visited on 26 July 2018, Xia and Wishart 2016) [25]. Data were normalized (sum), transformed (log), and scaled (mean-centered by SD of each variable).

3. Results and Discussion

3.1. Identification of the Metabolites

By performing ultra-high performance liquid chromatography quadrupole-time of flight mass spectrometry (UHPLC-Q/TOF) in negative ionization and the identification of metabolites using the grape and wine database *GrapeMetabolomics* [21], on average 350–400 compounds were putatively identified for each of the four grape varieties. The identity of the metabolites belonging to the chemical classes of flavonols/flavanones, glycoside terpenols (aroma precursors), flavanols and procyanidins, and stilbenes, was successively confirmed by multiple mass spectrometry (MS/MS), and their potential as wine varietal markers was evaluated.

Among them, a $[M − H]^-$ signal at m/z 285.068 corresponding to the molecular formula $C_{16}H_{13}O_5$, was observed in all the samples (mass error 1.4 ppm). This compound, eluting at 22.32 min, showed as main MS/MS fragments the signals at m/z 270.052 corresponding to the $C_{15}H_{10}O_5$ ion formed by •CH_3 loss (mass error −2.9 ppm), at m/z 243.066 corresponding to $C_{14}H_{11}O_4$ ion formed by CH_2CO

loss (mass error -0.8 ppm), and as mass spectrum base peak the signal at m/z 164.011 corresponding to the $C_8H_4O_4$ ion (mass error -2.4 ppm) (Figure 1). This compound was putatively identified as a methyl-naringenin isomer.

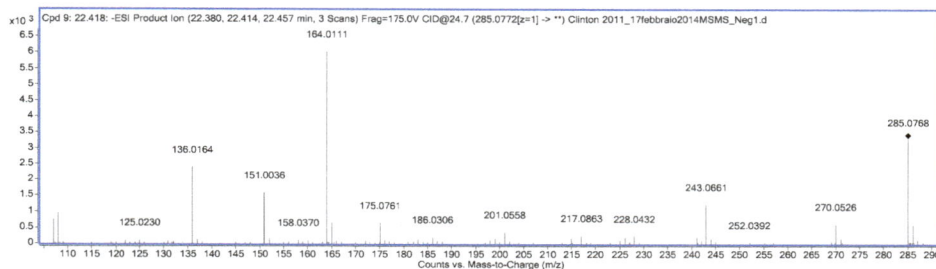

Figure 1. Ultra-high performance liquid chromatography quadrupole-time of flight multiple mass spectrometry (UHPLC-Q/TOF) spectrum of putative methyl-naringenin isomer identified in the grape.

Positive-MS analysis provided the identification of the grape anthocyanins, in particular delphinidin (Dp), cyanidin (Cy), petunidin (Pt), peonidin (Pn), and malvidin (Mv) glucoside, alongside with their acetylglucoside and *p*-coumaroylglucoside derivatives, and of Mv-caffeoylglucoside.

A total of 92 metabolites were identified in the samples, including 35 flavonols/flavanones, 16 anthocyanins, 11 glycoside monoterpenes, 11 flavanols/procyanidins, and 19 stilbenes. The potential for these metabolites to be used as a marker of the corresponding wines was then investigated.

3.2. Potential Flavonoid Markers of the Wine Varieties

Polyphenolic biosynthesis is regulated by genetic factors and several chemotaxonomic studies have shown that grape varieties can be differentiated on the basis of their anthocyanin and flavonol profiles [8,26–28]. In fact, despite that their amounts in grape are affected by environmental and agronomical factors, the profiles mainly depend on the cultivar characteristics [29]. In particular, while the F3′H enzyme is always active, the activity of the flavonoid 3′5′hydroxylase enzyme (F3′5H) varies depending on the grape variety [30]. Therefore, even if the phenolic parameters can be affected by the winemaking techniques and wine aging conditions used [6], anthocyanins and flavonols and their derivatives can be probably evaluated as potential variety traceability markers of wines [28,30–32].

Figure 2 shows the biplot of principal component analysis (PCA) calculated by using as variables the flavonols and flavanones identified in the grape varieties. Results indicate that the first two components account for 72.4% of the total variance, first component 37.0% and second component 35.4%.

The PCA clearly visualizes the separation among the varieties based on the non-anthocyanic flavonoids. The separation of the second component is driven by high contents of methyl-naringenin, myricetin (Mr), isorhamnetin (Iso), a tetrahydroxy-dimethoxy flavanone hexoside, three *p*-coumaroyl derivatives (kaempferide-*p*-coumaroylhexoside, isorhamnetin-*p*-coumaroylglucoside, and dihydrokaempferide-*p*-coumaroylhexoside) which were found in Raboso Piave. Tukey's test ($p < 0.01$) confirmed the statistical significance of these differences towards the other varieties (Table 1). Also statistically significant is the difference for the concentration of quercetin (Q) glucuronide, that was the lowest in Raboso Piave.

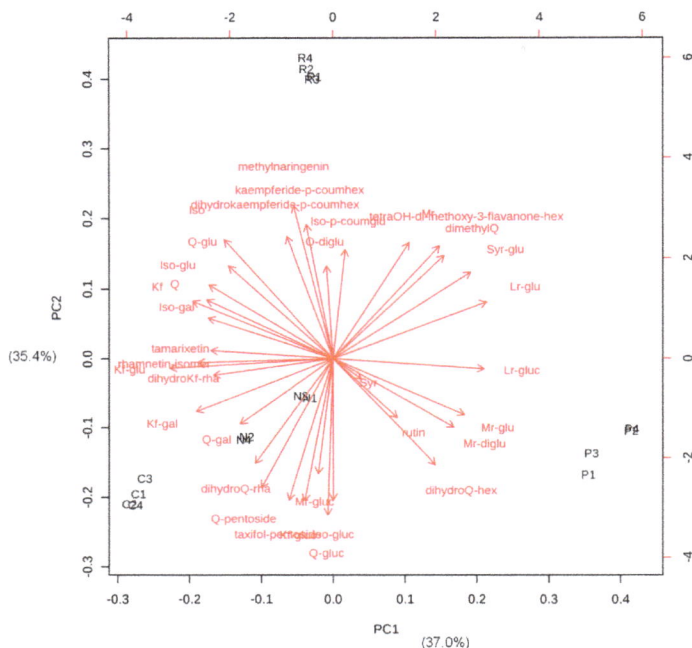

Figure 2. Biplot of the normalized UHPLC-Q/TOF signal intensities of non-anthocyanic flavonoids identified in the four grape varieties. P: Primitivo, C: Corvina, N: Negro Amaro, R: Raboso Piave. Mr, myricetin; Iso, isorhamnetin; Q, quercetin; Syr, syringetin; Lr, laricitrin; Kf, kaempferol; glu, glucoside; diglu, diglucoside; gluc, glucuronide; hex, hexoside; gal, galactoside; rha, rhamnoside; p-coumhex, *p*-coumaroylhexoside.

Primitivo grapes had contents of laricitrin (Lr) glucoside and Lr-glucuronide, as well as dihydroquercetin hexoside, that were significantly higher than in the other varieties.

The PCA also highlights in Primitivo high signals of Mr-glucoside and its diglucoside derivative, which are, however, not significantly different from those found in Negro Amaro. This variety also showed particularly low contents of tamarixetin, Iso, and kaempferol (Kf) derivatives.

Corvina is characterized by higher signals of taxifolin-pentoside and dihydroquercetin rhamnoside, the difference of which was statistically significant. As shown in Figure 2, lower signals of Lr and Syr derivates were observed in this variety.

Lastly, Negro Amaro showed significantly higher levels of Mr-glucuronide, Q- and Iso-galactosides, and tamarixetin (Table 1 and Figure 2).

Figure 3 shows the biplot of PCA of the four varieties calculated using the anthocyanins as variables. The first two components accounted for 78.5% of the total variance, with the first component 51.4% and the second component 27.1% of the variance. The four varieties were clearly separated also by their anthocyanin content. In particular, Raboso had significantly higher content of acetyl derivatives, in particular Dp, Cy, Pt, and Pn acetylglucosides, and Cy-*p*-coumaroylglucoside ($p < 0.01$, Tukey's test in Table 2). Significantly higher levels of Mv derivatives and Dp-*p*-coumaroylglucoside were found in Primitivo. Conversely, this variety had the lowest level of Cy-glucoside when compared to the other three varieties. In Corvina, a statistically significant low signal of Pt-glucoside was observed. Negro Amaro was mainly characterized by significantly higher Dp-glucoside levels, and the signals of acyl-anthocyanins had low intensities (as visualized by PCA), however they were not significantly different from the other varieties.

Table 1. Tukey's test calculated using the normalized UHPLC-Q/TOF signal intensities of flavonols and flavanones identified in the four grape varieties (*n* = 4). The data with different letters are significantly different for *p* < 0.01. n.f., signal not found.

Flavonols/Flavanones	*p* < 0.01			
	Corvina	Primitivo	Negro Amaro	Raboso
dihydrokaempferol-rhamnoside	b	a	a	c
dihydroquercetin-hexoside	a	c	ab	b
dihydroquercetin-rhamnoside	b	a	a	a
dimethylquercetin	a	b	a	c
isorhamnetin	a	n.f.	a	b
isorhamnetin-galactoside	a	b	c	a
isorhamnetin-glucoside	b	c	a	a
isorhamnetin-glucuronide	ab	ab	a	b
kaempferol	ab	a	ab	b
kaempferol-galactoside	a	b	ab	ab
kaempferol-glucoside	a	b	a	a
kaempferol-glucuronide	a	ab	ab	b
laricitrin-glucoside	a	b	a	c
laricitrin-glucuronide	a	c	b	ab
methylnaringenin	a	a	a	b
myricetin	a	b	c	d
myricetin-diglucoside	a	b	b	a
myricetin-glucoside	a	b	b	a
myricetin-glucuronide	a	a	b	a
quercetin	ab	b	c	ac
quercetin-diglucoside	a	a	b	b
quercetin-galactoside	a	b	c	ab
quercetin-glucoside	ab	a	bc	c
quercetin-glucuronide	a	b	a	c
quercetin-pentoside	a	a	a	a
rhamnetin-isomer	a	b	a	a
rutin	a	a	a	a
syringetin	a	ab	b	ab
syringetin-glucoside	b	a	c	a
tamarixetin	a	b	c	a
taxifolin-pentoside	b	a	a	a
tetrahydroxy-dimethoxyflavanone hexoside	b	a	a	c
kaempferide-*p*-coumaroylhexoside	a	a	a	b
isorhamnetin-*p*-coumaroylglucoside	a	a	b	c
dihydrokaempferide-*p*-coumaroylhexoside	a	a	a	b

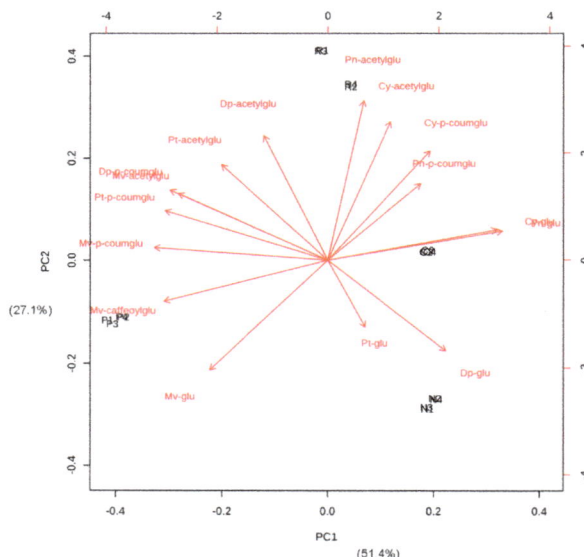

Figure 3. Biplot of the normalized UHPLC-Q/TOF anthocyanin signal intensities in the four grape varieties studied. P: Primitivo, C: Corvina, N: Negro Amaro, R: Raboso Piave. Dp, delphinidin; Cy, cyanidin; Pt, petunidin; Pn, peonidin; Mv, malvidin. glu, glucoside; p-coumglu, *p*-coumaroylglucoside; acetylglu, acetylglucoside.

Table 2. Tukey's test calculated using the normalized UHPLC-Q/TOF signal intensities of anthocyanins in the four grape varieties (*n* = 4). The data with different letters are significantly different for *p* < 0.01. Cy, cyanidin; Dp, delphinidin; Mv, malvidin; Pn, peonidin; Pt, petunidin.

Anthocyanins	*p* < 0.01			
	Corvina	**Primitivo**	**Negro Amaro**	**Raboso**
Cy-acetylglucoside	a	a	a	b
Cy-*p*-coumaroylglucoside	b	a	a	c
Cy-glucoside	b	c	a	a
Dp-acetylglucoside	a	a	a	b
Dp-*p*-coumaroylglucoside	a	b	a	c
Dp-glucoside	a	a	b	ab
Mv-acetylglucoside	a	b	a	b
Mv-caffeoylglucoside	a	b	a	a
Mv-*p*-coumaroylglucoside	a	b	a	c
Mv-glucoside	b	c	a	a
Pn-acetylglucoside	a	a	a	b
Pn-*p*-coumaroylglucoside	b	a	a	b
Pn-glucoside	a	c	ab	b
Pt-acetylglucoside	a	b	a	c
Pt-*p*-coumaroylglucoside	a	b	a	c
Pt-glucoside	a	ab	b	b

By performing Liquid chromatography coupled to high-resolution mass spectrometry (LC-Q/TOF) metabolomic analysis, also flavan-3-ols and procyanidins in pulp and skins, were identified. Table 3 reports the normalized signal intensities of flavan-3-ol monomers, dimers, and trimers identified in the berries of the samples after seeds had been removed. Corvina grapes had the highest procyanidin content, which was almost 4-fold higher than that of Raboso Piave and 2-fold than that of both Primitivo and Negro Amaro. Corvina also had the highest signals of (+)-catechin and procyanidin dimers. In a previous study, procyanidin B1 and B2 resulted determinant in discriminating the wines in terms of variety and origin [33].

Table 3. Normalized UHPLC-Q/TOF signal intensities of flavan-3-ols and procyanidin dimers and trimers identified in the berries removed by the seeds of the four grape varieties. CV%, coefficient of variance (SD × 100/mean, *n* = 4). In the last line, the percentages of total signal normalized to Raboso Piave are reported in bold. n.f., signal not found.

Procyanidins	Normalized [M − H]⁻ Signal Area							
	Corvina		Primitivo		Negro Amaro		Raboso	
	Mean	CV%	Mean	CV%	Mean	CV%	Mean	CV%
(−)-epicatechin	1,687,850	19	851,612	12	1,784,264	17	1,294,547	25
(+)-catechin	5,299,475	16	2,007,948	55	2,028,302	10	1,571,897	8
(−)-epigallocatechin	446,924	1	653,619	17	442,091	4	448,730	14
(−)-epicatechin gallate	984,163	24	445,224	25	729,303	21	51,256	14
procyanidin (B3/B4/B5)	1,314,453	15	477,508	8	882,147	18	289,062	2
procyanidin B1	6,813,192	4	2,778,443	7	2,501,480	8	839,493	6
procyanidin B2	111,031	15	n.f.		55,704	26	95,472	22
procyanidin T2/T3(T4)/C1	927,837	6	256,004	6	261,175	13	58,481	18
procyanidin T2/T3(T4)/C1	280,969	6	67,896	39	104,278	14	28,399	12
procyanidin T2/T3(T4)/C1	312,539	15	120,654	6	264,665	20	45,451	18
prodelphinidin T2/T3	134,478	10	51,808	26	89,307	9	41,249	17
Sum	18,312,911	(384%)	7,710,716	(162%)	9,142,716	(192%)	4,764,039	(100%)

In our study seeds were not analyzed, therefore their contribution to the wine procyanidin profile was not evaluated. Hence, these data just show the differences among the grape varieties but cannot be used for a wine traceability model.

In the biosynthesis of anthocyanins, the enzymes 3′methyltransferase (3′OMT) and flavonoid-3′,5′-hydroxylase (F3′5′H) transform Cy into Pn and into Dp, respectively. Higher F3′5′H activity increases the levels of trihydroxylated anthocyanins by affecting the dihydroxy/trihydroxy ratios, while 3′OMT induces methylation of Dp with formation of Pt and Mv [32,34]. A study on the F3′H and F3′5′H genes' expression showed a close relationship between the biosynthetic pathways of flavonols and anthocyanins [35].

With regard to our varieties, Primitivo grape is dominated by the presence of tri-substituted flavonoids, such as Lr, Mr, and Syr, as well as high content of tri-substituted anthocyanins, such as Pt and Mv derivatives. On the other hand, Corvina and Negro Amaro were found to be richer in di-substituted compounds. Raboso is characterized by a significant presence of both anthocyanic and non-anthocyanic acyl derivatives.

A study of Sangiovese wines showed that the wine anthocyanic pattern recognition is linked to the grape variety and the pigments formed during aging, such as vitisin B-like and vitisin A-like compounds, and ethyl-linked and direct-linked flavanol-anthocyanin derivatives [36]. The structures of these pigments are shown in Figure 4.

Taking these findings into consideration, one would expect to find in Primitivo wines higher amounts of the pigments formed by Pt and Mv, while in Raboso young wines, higher acyl anthocyanins are expected. Moreover, Primitivo and Raboso young wines can have significant *p*-coumaroyl anthocyanins, different from Negro Amaro, Corvina, or Sangiovese wines [36], and Raboso also high acetyl anthocyanins. However, the simple grape anthocyanins and their acyl derivatives that are usually present in large quantities in young wines, gradually decrease during aging due to degradation processes and reactions leading to the formation of more stable pigments. For example, vitisin A-like and vitisin B-like pigments are more stable than the corresponding grape anthocyanins [37], and Pinotin A-like pigments were found to increase with wine ageing [38].

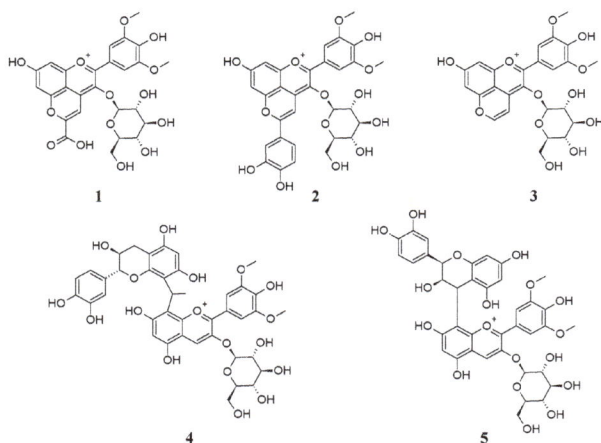

Figure 4. Pigments formed during wine aging: (**1**) vitisin A; (**2**) pinotin A; (**3**) vitisin B; (**4**) ethyl-linked catechin-Mv glucoside; (**5**) direct-linked catechin-Mv glucoside.

A study performed on Primitivo wines showed that the presence of Mv-*p*-coumaroylglucoside persists also in 2-year old wines [39]. The high Mv-glucoside content we found in Primitivo grapes indicated that aged wines probably have important content of Mv derivatives, such as pyranoanthocyanidins and flavanol-anthocyanin adducts. This assumption was confirmed by the study of Dipalmo et al., who identified the presence of many Mv-pigments in the 2-year old Primitivo wines, such as Mv-glucoside-4-vinyl-phenol, Mv-glucoside-4-vinyl-(epi)catechin, Mv-glucoside-8-ethyl-(epi)catechin, Mv-(*p*-coumaroyl)-glucoside-8-ethyl-(epi)catechin, (epi)-catechin -Mv-glucoside, di(epi)catechin-Mv-glucoside, Mv-acetylglucoside-4-vinyl-di(epi)catechin, Mv-(*p*-coum aroyl)-glucoside-4-vinyl-(epi)catechin, Mv-glucoside-8-ethyl-(epi)catechin, Mv-glucoside-4-vinyl-tri (epi)catechin, Mv-(caffeoyl)-glucoside-4-vinyl-di(epi)catechin, and Mv-(*p*-coumaroyl)-glucoside-4-vin yl-di(epi)catechin [39].

The high contents of Pn-glucoside and (+)-catechin found in Corvina grape suggest, during wine ageing, the formation of Pn-catechin derivatives which can be both direct-linked and ethyl-linked.

In general, flavonol and flavanone aglycones are present in wines as a result of the hydrolysis of corresponding glycosides occurring during winemaking [40]. Conversely, during wine aging, flavonols show different evolution patterns, a behavior that in some cases was observed and is dependent on the grape variety studied [41]. A study on red wines stabilized for 5 months showed a significant decrease of glycoside flavonols as result of the sugar moiety hydrolysis, and a significant decrease of total flavonol content due to their oxidation and co-pigmentation with anthocyanins [42,43].

It can be hypothesized that Primitivo grapes, characterized by high tri-substituted flavonols, produce wines richer in Lr, Mr, and Syr (aglycones or glycosides). In previous studies, the ratios between the total content of single flavonols were used to differentiate wine varieties. For example, the Q/Mr ratio was used to distinguish between Carménère and Merlot wines [44]. In Primitivo and Raboso wines, higher Mr/Q and Mr/Kf ratios are expected, being driven by the higher Mr and the lower Q and Kf in grapes. Primitivo wines can be also characterized by high Lr/Q and Lr/Kf ratios. On the contrary, lower Mr/Q and Mr/Kf ratios are expected in Corvina wines, being this variety characterized by lower Mr, Lr, and Syr and higher Kf and Q.

The abundance of Q and Kf glucuronides and galactosides, Q and taxifolin pentosides, and dihydroquercetin-rhamnoside could characterize the Negro Amaro and Corvina wines.

3.3. Monoterpene Glycosides (Aroma Precursors)

Wine aroma can be influenced by many factors, such as grape variety, climate, fermentation condition, yeast strains, winemaking process, aging, and storage conditions [45–47].

Glycoside monoterpenols are precursors responsible, in particular, for the aroma of aromatic and semi-aromatic grapes, e.g., Muscat and Malvasia varieties, Glera, Riesling, etc. Study of these secondary metabolites is also performed for grape chemotaxonomy aims [10,11,22,48], and wines from different varieties have been successfully differentiated on the basis of their terpene contents (e.g., nerol, β-santalol, 4-carene) [49].

A PCA performed using the monoterpene glycosides identified in the four grape varieties as variables, is shown in biplot Figure 5. The first two components accounted for 85.7% of the total variance, with the first component being 60.6% and the second component 25.1%. Results of the Tukey's test ($p < 0.01$) are reported in Table 4.

Table 4. Tukey's test calculated using the normalized UHPLC-Q/TOF signal intensities of monoterpene glycosides identified in the four grape varieties ($n = 4$). The data with different letters are significantly different for $p < 0.01$. n.f., signal not found.

Monoterpene Glycosides	$p < 0.01$			
	Corvina	Primitivo	Negro Amaro	Raboso
α-terpineol pentosyl-hexoside	n.f.	n.f.	a	b
linalool pentosyl-hexoside	n.f.	n.f.	a	b
geraniol pentosyl-hexoside	b	c	a	a
Ho-diendiol I pentosyl-hexoside	b	c	a	a
Ho-diendiol I rhamnosyl-hexoside	b	c	a	a
trans/cis 8-hydroxylinalool pentosyl-hexoside	a	a	b	c
trans/cis furan/pyran linalool oxide pentosyl-hexoside	a	a	b	n.f.
3,7-dimethyl-1-octen-6-one-3,7-diol pentosyl-hexoside 1	a	n.f.	a	b
3,7-dimethyl-1-octen-6-one-3,7-diol pentosyl-hexoside 2	a	a	a	b
3,7-dimethyl-1-octen-6-one-3,7-diol rhamnosyl-hexoside 1	a	a	b	c
3,7-dimethyl-1-octen-6-one-3,7-diol rhamnosyl-hexoside 2	b	a	c	a

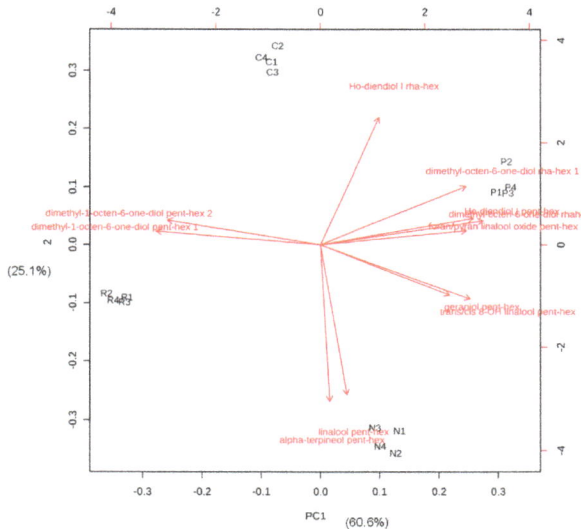

Figure 5. Biplot of the normalized UHPLC-Q/TOF signal intensities of monoterpene glycosides identified in the four varieties studied. P: Primitivo, C: Corvina, N: Negro Amaro, R: Raboso Piave. Pent-hex, pentosyl-hexoside; rha-hex, rhamnosyl-hexoside.

As observed for the anthocyanins and flavonols, the profiles of monoterpenol glycosides discriminate the four grape varieties. The separation along the second component was mainly driven by the high signals of linalool and α-terpineol pentosyl-hexoses, which were significantly lower in Raboso and were not detected in Corvina and Primitivo. Also, geraniol pentosyl-hexoside signal was very low in Corvina and Primitivo in respect to the other varieties. Negro Amaro and Raboso showed also a higher content of the Ho-diendiol I glycosides, and Raboso had a statistically significant high content of 3,7-dimethyl-1-octen-6-one-3,7-diol pentosyl-hexoses (Table 4). The high content of monoterpene glycosides found in Negro Amaro is in agreement with previous studies [50].

3.4. Other Metabolites

In addition to the compounds discussed above, the profiles of stilbenes in the four samples were detected and the normalized signal intensities are reported in Table 5.

Table 5. Normalized LC-Q/TOF signal intensities of resveratrol derivatives identified in the grape varieties studied. CV%, coefficient of variance (SD $\times$ 100/mean, $n = 4$). In the last line, the percentages of the total signal normalized to Raboso Piave samples are reported in bold. n.f., signal not found.

Stilbenes	Normalized [M − H]⁻ Signal Area							
	Corvina		Primitivo		Negro Amaro		Raboso	
	Mean	CV%	Mean	CV%	Mean	CV%	Mean	CV%
trans-resveratrol	283,900	53	22,549	17	59,096	10	807,651	15
piceatannol	269,846	45	106,655	49	205,075	30	1,400,281	4
cis-piceid	1,270,082	10	429,127	9	1,026,866	19	1,691,503	8
trans-piceid	177,040	17	99,308	14	96,404	30	313,008	15
E-astringin	67,013	15	47,798	10	68,551	9	44,347	10
Z-astringin	34,428	15	29,354	41	38,666	25	45,557	15
pallidol	246,754	8	71,397	30	42,236	26	172,312	18
resveratrol dimer 2	60,835	12	26,747	21	13,466	30	265,081	24
Z-ε-viniferin	169,259	2	63,423	15	72,224	24	1,771,218	18
E-ε-viniferin	187,388	10	110,183	13	68,958	4	990,456	14
Z-ω-viniferin	66,891	3	40,938	14	28,140	18	703,006	29
δ-viniferin	70,332	5	10,490	17	23,925	14	137,580	41
caraphenol	22,200	25	11,504	23	4690	10	126,287	43
pallidol-3-O-glucoside	33,095	12	29,203	9	14,455	12	91,436	11
α-viniferin	7699	53	59,992	37	16,784	83	113,833	57
Z-miyabenol C	25,288	4	18,977	17	11,579	21	342,147	23
E-miyabenol C	62,101	25	118,412	24	39,757	24	1,256,874	35
tetramer resveratrol 1	60,484	65	11,364	50	8958	14	152,028	49
tetramer resveratrol 2	31,491	27	12,509	48	8728	25	1,129,196	18
Sum	3,146,127	(27%)	1,319,930	(11%)	1,848,558	(16%)	11,553,801	(100%)

Several differences among the samples were found. In particular, the total signal of stilbenes in Raboso was up to 1–2 magnitude order higher than the other samples, *trans*-resveratrol was over 30-fold than Primitivo and 10-fold than Negro Amaro. A similar trend was also observed for piceatannol and the resveratrol oligomers.

Stilbenes accumulation in grape is induced by genetic factors, but viniferins and resveratrol oligomers are phytoalexins which can be synthetized as "inducible" compounds through the activation of the stilbene synthase gene (STS) under the elicitation of biotic and/or abiotic agents [51,52]. As a consequence, these compounds can hardly be considered as pure variety markers and were not evaluated for wine traceability in this study.

4. Conclusions

LC-Q/TOF *suspect screening analysis* provided the identification and relative quantification of metabolites belonging to the main chemical classes in the four grape varieties. This grape chemotaxonomy approach allowed the identification of several potential variety markers, which are likely to be found also in the resulting wines.

In Raboso wines, relatively high Mr/Q and Mr/Kf ratios (around 1 and 4, respectively) and a high abundance of both anthocyanic and non-anthocyanic acyl derivatives (in particular acetyl anthocyanins in young wines), are expected. The volatile fraction of these wines is probably characterized by the presence of 3,7-dimethyl-1-octen-6-one-3,7-diol and Ho-diendiol I.

Primitivo wines potentially have high contents of tri-substituted flavonoids, such as Lr, Mr, and Syr, and lower Iso and Kf derivatives. High Mr/Q and Mr/Kf ratios (around 1 and 6, respectively) and relatively high Lr/Q and Lr/Kf ratios (0.1 and 0.3, respectively), are expected. Wine color is characterized by high Pt and Mv pigments, with a significant presence of *p*-coumaroyl anthocyanins in young wines, and Pt and Mv pyranoanthocyanidins and flavanol-anthocyanin adducts in aged wines.

In general, Corvina wines are likely to have higher level of di-substituted compounds and lower acyl derivatives, with significant presence of taxifolin and dihydroquercetin, and low Lr and Syr. Young wines can be characterized by the presence of Q and Kf glucuronides and galactosides, Q and taxifolin pentosides and dihydroquercetin-rhamnoside, and low Mr/Q and Mr/Kf ratios (around 0.2 and 1, respectively). In aged wines, the presence of Pn-flavanol derivatives can be expected.

Negro Amaro wines have a non-anthocyanic flavonoid profile similar to Corvina, with higher di-substituted compounds, lower acyl derivatives, and a significant presence of Q, Kf, taxifolin, and dihydroquercetin. The volatile fraction will likely present peculiarly high levels of monoterpenols, such as α-terpineol, linalool, geraniol, and Ho-diendiol I.

It is worthy to note that the samples studied were collected from the same vine collection in just one vintage. Consequently, these findings do not take into account key variables such as vineyard location and vintage. However, this approach can potentially be applied to different study models and other high-quality wines. Despite the alcoholic fermentation impacts on the metabolites profile of a wine, generally the products partially maintain the varietal profiles. By comparing our findings and the previous results, the traceability markers here proposed can be probably applied to the wines. Future studies conducted on wines can confirm the hypotheses proposed.

Author Contributions: Conceptualization, R.F.; Methodology, M.D.R and A.D.V.; Software, M.D.R. and C.M.; Writing-Original Draft Preparation, R.F. and C.M.; Writing-Review & Editing, R.F. and C.M.; Visualization, M.R. and C.M.; Supervision, R.F.; Project Administration, R.F.

Funding: This research received no external funding.

Conflicts of Interest: The authors declare no conflict of interest.

References

1. De Rosso, M.; Panighel, A.; Carraro, R.; Padoan, E.; Favaro, A.; Della Vedova, A.; Flamini, R. Chemical characterization and enological potential of raboso varieties by study of secondary grape metabolites. *J. Agric. Food Chem.* **2010**, *2005*, 11364–11371. [CrossRef] [PubMed]
2. Del Gaudio, S.; Nico, G. Primitivo. In *Principali Vitigni da vino Coltivati in Italia*; Ministero dell'Agricoltura e delle Foreste: Roma, Italy, 1960.
3. Del Gaudio, S.; Panzera, C. Negro Amaro. In *Principali Vitigni da vino Coltivati in Italia*; Ministero dell'Agricoltura e delle Foreste: Roma, Italy, 1960.
4. Ragusa, A.; Centonze, C.; Grasso, M.E.; Latronico, M.F.; Mastrangelo, P.F.; Sparascio, F.; Fanizzi, F.P.; Maffia, M. A comparative study of phenols in apulian Italian wines. *Foods* **2017**, *6*, 24. [CrossRef] [PubMed]
5. Holmberg, L. Wine fraud. *Int. J. Wine Res.* **2010**, *2*, 105–113. [CrossRef]
6. Versari, A.; Laurie, V.F.; Ricci, A.; Laghi, L.; Parpinello, G.P. Progress in authentication, typification and traceability of grapes and wines by chemometric approaches. *Food Res. Int.* **2014**, *60*, 2–18. [CrossRef]
7. Villano, C.; Lisanti, M.T.; Gambuti, A.; Vecchio, R.; Moio, L.; Frusciante, L.; Aversano, R.; Carputo, D. Wine varietal authentication based on phenolics, volatiles and DNA markers: State of the art, perspectives and drawbacks Clizia. *Food Control* **2017**, *80*, 1–10. [CrossRef]
8. Mattivi, F.; Guzzon, R.; Vrhovsek, U.; Stefanini, M.; Velasco, R. Metabolite profiling of grape: Flavonols and anthocyanins. *J. Agric. Food Chem.* **2006**, *54*, 7692–7702. [CrossRef] [PubMed]

9. Figueiredo-González, M.; Martínez-Carballo, E.; Cancho-Grande, B.; Santiago, J.L.; Martínez, M.C.; Simal-Gándara, J. Pattern recognition of three *Vitis vinifera* L. red grapes varieties based on anthocyanin and flavonol profiles, with correlations between their biosynthesis pathways. *Food Chem.* **2012**, *130*, 9–19. [CrossRef]

10. Ghaste, M.; Narduzzi, L.; Carlin, S.; Vrhovsek, U.; Shulaev, V.; Mattivi, F. Chemical composition of volatile aroma metabolites and their glycosylated precursors that can uniquely differentiate individual grape cultivars. *Food Chem.* **2015**, *188*, 309–319. [CrossRef] [PubMed]

11. Nasi, A.; Ferranti, P.; Amato, S.; Chianese, L. Identification of free and bound volatile compounds as typicalness and authenticity markers of non-aromatic grapes and wines through a combined use of mass spectrometric techniques. *Food Chem.* **2008**, *110*, 762–768. [CrossRef]

12. Favretto, D.; Flamini, R. Application of electrospray ionization mass spectrometry to the study of grape anthocyanins. *Am. J. Enol. Vitic.* **2000**, *51*, 55–64.

13. Flamini, R.; Della Vedova, A.; Calo, A. Study of the monoterpene contents of 23 accessions of Muscat grape: Correlation between aroma profile and variety. *Riv. Vitic. Enol.* **2001**, *54*, 35–50.

14. De Rosso, M.; Tonidandel, L.; Larcher, R.; Nicolini, G.; Ruggeri, V.; Dalla Vedova, A.; De Marchi, F.; Gardiman, M.; Flamini, R. Study of anthocyanic profiles of twenty-one hybrid grape varieties by liquid chromatography and precursor-ion mass spectrometry. *Anal. Chim. Acta* **2012**, *732*, 120–129. [CrossRef] [PubMed]

15. Rubert, J.; Lacina, O.; Fauhl-hassek, C.; Hajslova, J. Metabolic fingerprinting based on high-resolution tandem mass spectrometry: A reliable tool for wine authentication? *Anal. Bioanal. Chem.* **2014**, *406*, 6791–6803. [CrossRef] [PubMed]

16. Vaclavik, L.; Lacina, O.; Hajslova, J.; Zweigenbaum, J. The use of high performance liquid chromatography-quadrupole time-of-flight mass spectrometry coupled to advanced data mining and chemometric tools for discrimination and classification of red wines according to their variety. *Anal. Chim. Acta* **2011**, *685*, 45–51. [CrossRef] [PubMed]

17. Arapitsas, P.; Ugliano, M.; Perenzoni, D.; Angeli, A.; Pangrazzi, P.; Mattivi, F. Wine metabolomics reveals new sulfonated products in bottled white wines, promoted by small amounts of oxygen. *J. Chromatogr. A* **2016**, *1429*, 155–165. [CrossRef] [PubMed]

18. Arbulu, M.; Sampedro, M.C.; Gómez-caballero, A.; Goicolea, M.A.; Barrio, R.J. Untargeted metabolomic analysis using liquid chromatography quadrupole time-of-flight mass spectrometry for non-volatile profiling of wines. *Anal. Chim. Acta* **2015**, *858*, 32–41. [CrossRef] [PubMed]

19. Arapitsas, P.; Della Corte, A.; Gika, H.; Narduzzi, L.; Mattivi, F. Studying the effect of storage conditions on the metabolite content of red wine using HILIC LC–MS based metabolomics. *Food Chem.* **2016**, *197*, 1331–1340. [CrossRef] [PubMed]

20. Flamini, R.; De Rosso, M.; Bavaresco, L. Study of grape polyphenols by liquid chromatography-high-resolution mass spectrometry (UHPLC/QTOF) and suspect screening analysis. *J. Anal. Methods Chem.* **2015**, *2015*. [CrossRef] [PubMed]

21. Flamini, R.; De Rosso, M.; De Marchi, F.; Dalla Vedova, A.; Panighel, A.; Gardiman, M.; Maoz, I.; Bavaresco, L. An innovative approach to grape metabolomics: Stilbene profiling by suspect screening analysis. *Metabolomics* **2013**, *9*, 1243–1253. [CrossRef]

22. Flamini, R.; De Rosso, M.; Panighel, A.; Dalla Vedova, A.; De Marchi, F.; Bavaresco, L. Profiling of grape monoterpene glycosides (aroma precursors) by ultra-high performanceliquid chromatography-high resolution mass spectrometry (UHPLC/QTOF). *J. Mass Spectrom.* **2014**, *49*, 1214–1222. [CrossRef] [PubMed]

23. Di Stefano, R.; Flamini, R. *High Performance Liquid Chromotagraphy Analysis of Grape and Wine Polyphenols*; John Wiley & Sons: Hoboken, NJ, USA, 2008; pp. 33–79.

24. Pezet, R.; Perret, C.; Jean-Denis, J.B.; Tabacchi, R.; Gindro, K.; Viret, O. δ-Viniferin, a resveratrol dehydrodimer: One of the major stilbenes synthesized by stressed grapevine leaves. *J. Agric. Food Chem.* **2003**, *51*, 5488–5492. [CrossRef] [PubMed]

25. Xia, J.; Wishart, D.S. Using MetaboAnalyst3.0 for comprehensive metabolomics data analysis. *Curr. Protoc. Bioinform.* **2016**, *55*, 14.10.1–14.10.91. [CrossRef] [PubMed]

26. Fernández-López, J.A.; Almela, L.; Muñoz, J.A.; Hidalgo, V.; Carreño, J. Dependence between colour and individual anthocyanin content in ripening grapes. *Food Res. Int.* **1998**, *31*, 667–672. [CrossRef]

27. Figueiredo-Gonzalez, M.; Cancho-Grande, B.; Simal-Gandara, J. Evolution of colour and phenolic compounds during Garnacha Tintorera grape raisining. *Food Chem.* **2013**, *141*, 3230–3240. [CrossRef] [PubMed]

28. Ortega-Regules, A.; Romero-Cascales, I.; López-Roca, J.M.; Ros-García, J.M.; Gómez-Plaza, E. Anthocyanin fingerprint of grapes: Environmental and genetic variations. *J. Sci. Food Agric.* **2006**, *86*, 1460–1467. [CrossRef]

29. Liang, N.-N.; Zhu, B.-Q.; Han, S.; Wang, J.-H.; Pan, Q.-H.; Reeves, M.J.; Duan, C.-Q.; He, F. Regional characteristics of anthocyanin and flavonol compounds from grapes of four *Vitis vinifera* varieties in five wine regions of China. *Food Res. Int.* **2014**, *64*, 264–274. [CrossRef] [PubMed]

30. Squadrito, M.; Corona, O.; Ansaldi, G.; Di Stefano, R. Possible relations between biosynthetic pathways of HCTA, flavonols and anthocyanins in grape berry skin. *Riv. Vitic. Enol.* **2007**, *60*, 59–70.

31. Cacho, J.; Fernández, P.; Ferreira, V.; Castells, J.E. Evolution of five anthocyanidin-3-glucosides in the skin of the Tempranillo, Moristel, and Garnacha grape varieties and influence of climatological variables. *Am. J. Enol. Vitic.* **1992**, *43*, 244–248.

32. Carreño, J.; Almela, L.; Martínez, A.; Fernández-López, J.A. Chemotaxonomical classification of red table grapes based on anthocyanin profile and external colour. *LWT Food Sci. Technol.* **1997**, *30*, 259–265. [CrossRef]

33. Makris, D.P.; Kallithraka, S.; Mamalos, A. Differentiation of young red wines based on cultivar and geographical origin with application of chemometrics of principal polyphenolic constituents. *Talanta* **2006**, *70*, 1143–1152. [CrossRef] [PubMed]

34. Heller, W.; Forkmann, G. Biosynthesis. In *The Flavonoids: Advances in Research since 1980*; Chapman & Hall: London, UK, 1980; pp. 399–425.

35. Jeong, S.T.; Goto-Yamamoto, N.; Hashizume, K.; Esaka, M. Expression of the flavonoid 3′-hydroxylase and flavonoid 3′,5′-hydroxylase genes and flavonoid composition in grape (*Vitis vinifera*). *Plant Sci.* **2006**, *170*, 61–69. [CrossRef]

36. Arapitsas, P.; Perenzoni, D.; Nicolini, G.; Mattivi, F. Study of Sangiovese Wines pigment profile by UHPLC-MS/MS. *J. Agric. Food Chem.* **2012**, *60*, 10461–10471. [CrossRef] [PubMed]

37. Fulcrand, H.; Dueñas, M.; Salas, E.; Cheynier, V. Phenolic reactions during winemaking and aging. *Am. J. Enol. Vitic.* **2006**, *57*, 289–297.

38. Rentzsch, M.; Schwarz, M.; Winterhalter, P.; Hermosín-Gutiérrez, I. Formation of hydroxyphenyl-pyranoanthocyanins in Grenache wines: Precursor levels and evolution during aging. *J. Agric. Food Chem.* **2007**, *55*, 4883–4888. [CrossRef] [PubMed]

39. Dipalmo, T.; Crupi, P.; Pati, S.; Lisa, M.; Di, A. Studying the evolution of anthocyanin-derived pigments in a typical red wine of Southern Italy to assess its resistance to aging. *LWT Food Sci. Technol.* **2016**, *71*, 1–9. [CrossRef]

40. Castillo-Muñoz, N.; Gómez-Alonso, S.; García-Romero, E.; Hermosín-Gutiérrez, I. Flavonol profiles of *Vitis vinifera* red grapes and their single-cultivar wines. *J. Agric. Food Chem.* **2007**, *55*, 992–1002. [CrossRef] [PubMed]

41. Monagas, M.; Gómez-Cordovés, C.; Bartolomé, B. Evolution of polyphenols in red wines from *Vitis vinifera* L. during aging in the bottle I. Anthocyanins and pyranoanthocyanins. *Eur. Food Res. Technol.* **2005**, *220*, 607–614. [CrossRef]

42. Lingua, M.S.; Fabani, M.P.; Wunderlin, D.A.; Baroni, M.V. From grape to wine: Changes in phenolic composition and its influence on antioxidant activity. *Food Chem.* **2016**, *208*, 228–238. [CrossRef] [PubMed]

43. Bimpilas, A.; Tsimogiannis, D.; Balta-brouma, K.; Lymperopoulou, T.; Oreopoulou, V. Evolution of phenolic compounds and metal content of wine during alcoholic fermentation and storage. *Food Chem.* **2015**, *178*, 164–171. [CrossRef] [PubMed]

44. Von Baer, D.; Mardones, C.; Gutierrez, L.; Hofmann, G.; Becerra, J.; Hitschfeld, A.; Vergara, C. Varietal authenticity verification of Cabernet sauvignon, Merlot and Carmenere wines produced in Chile by their anthocyanin, flavonol and shikimic acid profiles. *Bull. OIV* **2005**, *887*, 45–57.

45. Cordente, A.G.; Curtin, C.D.; Varela, C.; Pretorius, I.S. Flavour-active wine yeasts. *Appl. Microbiol. Biotechnol.* **2012**, *96*, 601–618. [CrossRef] [PubMed]

46. Ebeler, S.E.; Thorngate, J.H. Wine chemistry and flavor: Looking into the crystal glass. *J. Agric. Food Chem.* **2009**, *57*, 8098–8108. [CrossRef] [PubMed]

47. Styger, G.; Prior, B.; Bauer, F.F. Wine flavor and aroma. *J. Ind. Microbiol. Biotechnol.* **2011**, 1145–1159. [CrossRef] [PubMed]

48. Flamini, R.; Menicatti, M.; De Rosso, M.; Gardiman, M.; Mayr, C.; Pallecchi, M.; Danza, G.; Bartolucci, G. Combining liquid chromatography and tandem mass spectrometry approaches to the study of monoterpene glycosides (aroma precursors) in wine grape. *J. Mass Spectrom.* **2018**, *53*, 1–35. [CrossRef] [PubMed]

49. Welke, J.E.; Manfroi, V.; Zanus, M.; Lazzarotto, M.; Alcaraz Zini, C. Differentiation of wines according to grape variety using multivariate analysis of comprehensive two-dimensional gas chromatography with time-of-flight mass spectrometric detection data. *Food Chem.* **2013**, *141*, 3897–3905. [CrossRef] [PubMed]

50. Tamborra, P.; Esti, M. Authenticity markers in Aglianico, Uva di Troia, Negroamaro and Primitivo grapes. *Anal. Chim. Acta* **2010**, *660*, 221–226. [CrossRef] [PubMed]

51. Goetz, G.; Fkyerat, A.; Metais, N.; Kunz, M.; Tabacchi, R.; Pezet, R.; Pont, V. Resistance factors to grey mould in grape berries: Identification of some phenolics inhibitors of Botrytis cinerea stilbene oxidase. *Phytochemistry* **1999**, *52*, 759–767. [CrossRef]

52. Bavaresco, L.; Mattivi, F.; De Rosso, M.; Flamini, R. Effects of elicitors, viticultural factors, and enological practices on resveratrol and stilbenes in grapevine and wine. *Mini Rev. Med. Chem.* **2012**, *12*, 1366–1381. [PubMed]

beverages

MDPI

Article

Influence of Chemical and Physical Variables on ^{87}Sr/^{86}Sr Isotope Ratios Determination for Geographical Traceability Studies in the Oenological Food Chain

Simona Sighinolfi [1], Caterina Durante [2], Lancellotti Lisa [1], Lorenzo Tassi [1] and Andrea Marchetti [1,*]

[1] Department of Geological and Chemical Science, University of Modena and Reggio Emilia, via Campi 103, 41125 Modena, Italy; simona.sighinolfi@unimore.it (S.S.); lisa.lancellotti@unimore.it (L.L.); lorenzo.tassi@unimore.it (L.T.)

[2] ChemStamp s.r.l., Spin off University of Modena and Reggio Emilia, via Campi 103, 41125 Modena, Italy; durante.caterina@gmail.com

* Correspondence: andrea.marchetti@unimore.it; Tel.: +39-059-205-8637

Received: 18 June 2018; Accepted: 24 July 2018; Published: 1 August 2018

Abstract: This study summarizes the results obtained from a systematic and long-term project aimed at the development of tools to assess the provenance of food in the oenological sector. ^{87}Sr/^{86}Sr isotope ratios were measured on a representative set of soils, branches, and wines sampled from the Chianti Classico wine production area. In particular, owing to the high spatial resolution of the ^{87}Sr/^{86}Sr ratio in the topsoil, the effect of two mill techniques for soil pretreatment was investigated to verify the influence of the particle dimension on the measured isotopic ratios. Samples with particle sizes ranging from 250 to less than 50 µm were investigated, and the extraction was performed by means of the DIN 19730 procedure. For each sample, the Sr isotope ratio was determined as well. The obtained results showed that the ^{87}Sr/^{86}Sr ratio is not influenced by soil particle size and may represent an effective tool as a geographic provenance indicator for the investigated product.

Keywords: geographical traceability; ^{87}Sr/^{86}Sr isotopic ratio; Chianti Classico wine; soil particle size

1. Introduction

Among the different criteria adopted to promote and protect food quality, the European Union (EU) has introduced, with the use of quality schemes, the link between territory and food [1]. In this context, the protected designation of origin (PDO) represents the highest award that can be attributed to an aliment, and it implies that the entire food chain is within a delimited territory [2]. Although there are many paper certifications for each food chain process that state for authenticity and quality arising from a particular geographical origin, none of these is based on objective criteria. In recent years, several attempts to develop tracking and tracing models for food processes have been made [3–5].

When dealing with traceability models, it is of utmost importance to determine the identity of the geographical indicator that is used to monitor the food chain from the field to the final product. For geographical traceability issues, one of the indicators that can be used is the primary or direct type, such as metal content or isotope ratios of bio-elements (^{13}C/^{12}C, D/H, ^{18}O/^{16}O, ^{15}N/^{14}N, and ^{34}S/^{32}S) or radiogenic heavy elements (^{87}Sr/^{86}Sr, ^{145}Nd/^{143}Nd, and ^{207}Pb/^{206}Pb). One element, whose isotopic pattern shows promising perspective in different areas, with particular and peculiar applications in food traceability, is strontium [5–15].

As a geographic food tracer, the most important feature of Sr is that this element is assimilated by the plant roots, and it has been shown that the isotopic ratio fingerprint, from the soil to the final product, remains almost unaltered (i.e., the isotopes do not undergo appreciable mass-dependent fractionation processes [16,17]). For a statistical approach to the geographical origin of food, a large dataset of precise and accurate values is needed in order to evaluate the indicator variability range of both the food and the soils and to build robust classification models [18].

If, on the one hand, it is a quite simple task to collect isotopic ratios data for the food matrix, on the other hand, the interpretation of the data from the soil is complex [19–21]. In fact, depending on the geology of the territory, the $^{87}Sr/^{86}Sr$ ratio could vary considerably as a function of many variables difficult to evaluate a priori [22,23]. Probably, the discrepancy between the bioavailable fraction and the total amount of strontium in a soil could represent the limiting factor for the construction of reliable geographical traceability models for food.

The possibility of discriminating between samples' origins for geographical traceability purposes often lies at the 4th or 5th decimal places of this isotopic ratio. Therefore, accuracy of the measured values represents a quality parameter of utmost importance to ascertain the goodness of the experimental results. In the case of isotopic ratio determinations, it must be stressed that precision and accuracy depend upon several factors. In the case of inductively coupled plasma mass spectrometry (ICP/MS), the determination of the data precision is mainly influenced by the instrumental setup and the measuring conditions [24] primarily related to the sample introduction system and aerosol formation, while accuracy is more dependent on non-spectroscopic and spectroscopic interferences [9,25,26]. In the case of $^{87}Sr/^{86}Sr$ determination, the isobaric interference of charged ions and mass bias phenomena, due to the different mass transmissions of the multiple ion beams produced in the plasma source, have to be considered and corrected as well [24].

In particular, the isobaric interferences are due to the presence of doubly-charged ions ($^{168}Yb^{2+}$, $^{168}Eu^{2+}$, $^{172}Yb^{2+}$, $^{174}Yb^{2+}$, etc.), charged polyatomic adducts ($^{42}Ca^{42}Ca^+$, $^{44}Ca^{40}Ar^+$, $^{46}Ca^{40}Ar^+$, $^{44}Ca^{44}Ca^+$, etc.), and other isotopes with the same m/z ratio, such as $^{87}Rb^+$ and $^{86}Kr^+$, which must be removed by chemical procedure for ^{87}Rb or mathematically corrected in the cases of ^{86}Kr and ^{87}Rb [27].

The solid phase extraction (SPE) procedure adopted for the Rb/Sr separation is characterized by an analytical recovery close to 100%, but the eluted Sr solution could contain some Rb traces at the ng kg^{-1}–μg kg^{-1} level. While this is not a problem for thermal ionization mass spectrometry (TIMS) determinations [11], some major concerns arise for ICP-based determinations. Rb has two natural isotopes, ^{85}Rb with a relative abundance of 72.17% and ^{87}Rb, a radioactive nuclide with a half decay time of 4.88 × 10^{10} years to ^{87}Sr, with a relative abundance of 27.83%. In addition, the isobaric interference of ^{87}Rb on ^{87}Sr cannot be instrumentally resolved, because it would need a resolving power close to 300,000: a resolution value still unavailable in commercial ICP/MS spectrometers. For these reasons, the SPE process must be strictly controlled to maximize the recovery, and the residual Rb, if present, must be corrected for. However, the mathematical correction for Rb, in addition to the one for ^{86}Kr, may give results whose uncertainty is dependent on the initial concentration of the isobaric interfering, and when the ratio between the Sr and Rb concentrations is below 1000, the method could be unsuitable [27,28].

The residual ^{87}Rb is evaluated by measuring the signal for mass 85 (^{85}Rb) with the aim to improve the accuracy of the $^{87}Sr/^{86}Sr$ ratio; furthermore, the $^{87}Rb/^{85}Rb$ ratio is used for the mathematical Rb correction. For both mathematical procedures, the ^{86}Kr and ^{87}Rb corrections, the isotope ratios are also corrected for mass bias discrimination by the exponential law, assuming that the Kr and Rb mass discrimination factors are the same as that of Sr [19,27].

In order to investigate the performance of the mathematical procedures to correct for Rb interference, several Sr reference solutions at different concentrations were spiked with defined amounts of Rb, and the solutions were measured, without any chemical Rb separation, for the $^{87}Sr/^{86}Sr$ ratio.

In addition to the above-cited factors that may influence the accuracy of the instrumental measurements of the Sr isotopic ratio, the representativeness and the homogeneity of the sample have to be taken into consideration. In fact, the particle size plays a key role when dealing with soil and bioavailability. Therefore, a grinding procedure is often necessary to increase the surface-to-volume ratio. Moreover, the milling pretreatment may introduce some artifacts, that can alter chemical analysis results, if other parts of the soil sample, such as stones and rocks, are processed as well.

Taking into account the above considerations, the choice of the most appropriate milling equipment is of utmost importance to obtain reliable results.

In particular, to process soil samples, centrifugal and planetary ball mills are both suggested to reduce the sample size. However, because these techniques are based on different physical principles, the final sample properties may be influenced. Unlike the centrifugal mill, whose working principle is mainly based on an impact-shearing effect, the grinding process by planetary mill can be easily improved by adjusting the setup parameters to leave almost unaltered the hardest stony particles present in the soil samples.

Owing to these problems and in addition to the difficulties arising in the evaluation of the bioavailable fraction in the soil, as suggested by other researchers [29,30] and as reported in previous works [18], it should be a promising alternative to move from the "passive" soil collecting activity to an "active" one, where plants do directly the sampling [30–32].

In fact, plant roots have direct access to the bioavailable element reservoir in the soil, extending the sampling uptake to the neighboring area of the growing vine/plants. As a consequence, the $^{87}Sr/^{86}Sr$ data are the result of the influence of several processes, such as weathering reactions and contamination produced by fertilizers [33], pesticides, aerosol uptake, and so on.

The use of the plants as sampling devices allows the following: (i) a direct access to the bioavailable element fraction; (ii) the ability to make integration over a larger soil volume of element uptake; and (iii) the capacity to realize a simpler integration over time/seasons procedure.

In this paper, both aspects related to the influence of the soil size distribution on the isotopic data and the effectiveness of the Rb mathematical correction have been tackled. Furthermore, a preliminary study was carried out in order to verify whether the Sr isotopic ratio coming from vine branches can be a more distinctive geographical traceability tool for the Chianti Classico wine production areas, as other studies on the provenance have shown [10,18,22].

2. Materials and Methods

2.1. Reagents and Materials

Ultrapure 65% HNO_3 was obtained from analytical grade HNO_3 (Carlo Erba, Milan, Italy) by means of a SAVILLEX DST 1000 sub-boiling system (Savillex Corp., Eden Prairie, MN, USA).

High-purity water (ASTM TYPE I) was obtained by a Milli-Q Element system (Millipore, Milan, Italy).

Suprapure 1 M NH_4NO_3 and 30% analytical grade H_2O_2 were from Merk Millipore, Milan, Italy.

$SrCO_3$, NIST SRM 987 (NIST, Gaithersburg, MD, USA), with a generally accepted Sr isotopic value of 0.71026 ± 0.00002 (u = 2sd) [34] was used for the bracketing procedure, evaluation of the data precision, and preparation of the spiked Sr/Rb solutions. $RbNO_3$, trace metal basis 99.95% purity from Merck, was used to prepare the Sr/Rb spiked solutions.

ICP-multielement solution, IV-ICP-MS-71A, used for the determination of the Sr and Rb concentration, was from Inorganic Ventures, Christiansburg, VA, USA.

The Eichrom SR-B100-S (50–100 µm) Sr resin, used for Sr/Rb separation, was purchased from Eichrom Europe Laboratories, France. Preparation of the resin and the separation procedure were described in a previous work [19].

All Perfluoroalkoxy (PFA) bottles, tubes, and vessels, used for solution and sample preparation and storage, were firstly washed with heated 10% HNO_3 and then rinsed with high-purity water.

Standard solution and sample preparation was carried out by weight with a Mettler AE200 analytical balance (Mettler Toledo S.p.A, Milan, Italy) with ± 0.0001 g sensitivity.

All standards and samples were processed under a horizontal laminar flow hood, equipped with a HEPA filter, to prevent contamination phenomena.

2.2. Instrumentation

The soil and vine branch grinding procedure was performed by means of a Fritch, Pulverisette 14 model centrifugal mill (ECO Scientifica, Milan, Italy), equipped with a pure titan 12 ribs rotor and a 500 μm trapezoidal perforation sieve ring with a Teflon-coated collecting pan. The rotor speed was set at 16,000 rpm for both the sample matrices.

The soil samples were also grinded using a PM100 planetary mill (Retsch, FKV, Bergamo, Italy), equipped with a single 250 mL agata jar with two milling balls. Table S1 (in the Supplementary Materials) reports the milling parameters. The soil samples were processed twice, but only a sample fraction with a grain size greater than 250 μm was treated in the second run.

An automated vibratory sieve shaker, model AS200, supplied by Retsch (FKV, Bergamo, Italy) was used for size partition. The sieving stack was composed by seven solid stainless-steel sieve frames—315 μm, 250 μm, 180 μm, 125 μm, 90 μm, 60 μm, and 50 μm—and a collecting pan for the <50 μm sample size. The sample size fractionation was obtained by a throwing motion with angular momentum, and the setup parameters were as follow: amplitude = 70% with a total processing time of 5 min. All the samples were processed in dry sieving mode.

A commercial microwave oven for laboratory use, MarsX model (CEM Corp., Bergamo, Italy), equipped with Teflon® XP1500plus-type closed vessels, was used to digest the vine branch samples. The instrument was equipped with active probes on a reference vessel for temperature control, RTP-300 Plus, and pressure control, ESP-1500 Plus, during mineralization.

The Sr concentration in all the sample solutions was determined using an inductively coupled plasma interfaced to a quadrupolar mass analyzer, ICP/qMS, (XSeries II model, ThermoFisher Scientific, Bremen, Germany). The instrumental parameters and experimental setup conditions are reported in Tables S2–S4 (in the Supplementary Materials) for the microwave closed vessels technique (MW) and ICP/qMS instruments, respectively.

The strontium isotope ratio data ($^{87}Sr/^{86}Sr$) were acquired using a double-focusing multi-collector inductively coupled plasma mass spectrometer (MC-ICP/MS) (Neptune, ThermoFisher Scientific, Bremen, Germany). This spectrometer consisted of double-focusing, electrostatic, and magnetic sectors arranged in a forward Nier–Johnson geometry, and a Faraday cup multi-collector detector. The data acquisition was performed in low-resolution mode. The instrumental parameters are reported in a previous study [19]. The ion lens setting was daily tuned for maximum sensitivity and optimal flat-topped peak-shaped signal.

3. Experimental

3.1. Soil Sampling

Soil samples were collected in two different areas, labeled site A and site B, located in the Chianti Classico wine production zone of Tuscany (Italy). Table S5 (in the Supplementary Materials) reports the global positioning system, GPS, coordinates for both the sampling sites in the degree, minute, seconds (DMS) scale. The soil holes were dug by a single gauge auger set for hardly disturbed samples at a maximum sampling depth of 70 cm, roughly. The obtained cores were divided in two aliquots, up (UP) and down (DW), 30 cm length each, discarding the upper 10 cm length. The collected soils were properly preprocessed. In particular, the soil cores were broken up, and stones and plant debris were removed. The samples were air-dried for one week, successively stored in polystyrene bottles, and kept at room temperature. The number of samples is proportional to the extension of the investigated vineyards. A large-scale sampling procedure was applied for this study, and a soil/vine branch sample was taken each 5000 m^2.

In addition, to investigate the milling procedure effects, four soil samples (M1, M2, M3, and M4) from district of Modena were collected following the same operative procedure and used to test the influence of the soil particle size on the $^{87}Sr/^{86}Sr$ isotopic ratio.

3.2. Vine Branch Sampling

During the soil sampling, vine branches were also collected from sites A and B. In particular, samples, 10–20 cm long, were cut from plants growing near the soil sampling point. The number of sampled vine branches was equivalent to the soil samples. The samples were dried in an oven for 24 h at 105 °C, and then, they were cut to 1 cm length, stored in polystyrene bottles, and kept at room temperature.

3.3. Wine Sampling

Grapes harvested from sites A and B were processed separately by the producer and from the respective grape juices after malolactic fermentation had taken place. Two different batches of wine, namely wine_A and wine_B, were obtained.

4. Sample Processing

4.1. Soil

Different aliquots of the same air-dried soil were grounded by means of centrifugal and/or planetary mill. Successively, only M1–M4 soil powders were processed for particle size distribution.

The determination of the bioavailable Sr fraction in the soil was achieved by extraction with 1 M NH_4NO_3 solution, DIN 19730 [35]. The obtained eluate was filtered through a 0.2-μm pore-sized cellulose acetate membrane into a 30 mL PFA bottle. Before the Sr/Rb separation, the total Sr concentration was measured, and the sample was diluted to a final concentration close to 200 μg kg^{-1} with 8 M HNO_3, to work within the optimal instrumental conditions in terms of accuracy and precision of the isotopic measurements [19].

4.2. Vine Branches and Wines

The mineralization of vine branches was performed by the microwave closed vessels technique (MW) on a maximum sample size of 0.3 g, accurately weighted into the MW reaction vessels and then supplemented with 6 mL 65% HNO_3, 1 mL 30% H_2O_2, and 3 mL H_2O [36]. Washing cycles, with 6 mL 65% HNO_3 and 4 mL H_2O, were always performed between each sample mineralization cycle by using the same heating program. At the end of the sample mineralization procedure, colorless or pale yellow solutions were always obtained. Also, for these samples, the total Sr concentration was determined before the Sr/Rb separation in order to determine the optimal dilution factor.

The digestion of the wine samples was carried out by a simplified, validated protocol consisting of a low-temperature mineralization procedure [16]. Briefly, a sample aliquot of 5 mL of wine was added to 5 mL of 65% HNO_3, and the mixture was left to react for 12 h in PFA bottles at room temperature and atmospheric pressure. Before the Rb/Sr SPE separation, the mineralized solution was then measured for the total Sr content in order to dilute sample to a final 200 μg kg^{-1} Sr concentration.

4.3. Strontium/Rubidium Spiking Solutions

A set of 20 solutions with different concentrations of Sr and Rb were prepared to test the influence of the isobaric ^{87}Rb specie on the accuracy of the $^{87}Sr/^{86}Sr$ ratio determination. In particular, four solutions of $SrCO_3$, NIST 987, at 50, 100, 200, and 400 μg kg^{-1} Sr were spiked with increasing amounts of $RbNO_3$ from 0 to 200 μg kg^{-1} Rb final concentration. The measurements were performed directly on the solutions without the Rb/Sr separation in triplicate and randomized in three days to avoid systematic deviations.

4.4. Strontium Isotopic Ratio Determination on Real Samples

Before starting any measurement procedure, all the samples were processed to separate interfering species such as rubidium. The Sr/Rb separation procedure was optimized by means of an experimental design approach and fully described in previous works [16,37]. Briefly, it consisted of the following steps: (i) 1–2 mL of resin was loaded into the SPE column; (ii) the resin was washed with high-purity water and activated with 8 M HNO_3; (iii) an appropriate sample volume was loaded and the interferences eluted with 8 M HNO_3; and finally, (v) the recovery of Sr was accomplished by using high-purity water.

A [blank/sample/blank/standard/blank] bracketing sequence was always adopted for Sr isotope ratio measurements in order to check and correct for any instrumental drift. The average of the measured intensities of bracketing-blanks (4% HNO_3) was subtracted from the measured intensities of the respective standards (NIST SRM 987) or samples.

The strontium isotope ratios of the standards and samples were calculated according to the mathematical procedures, as explained in a previous work [19], considering an internal correction for the mass-dependent fractionation process effects induced by the plasma source by using as a normalizing factor the $^{88}Sr/^{86}Sr$ ratio equal to 8.3752 (according to the IUPAC technical report on the isotopic composition of the elements [38]).

Furthermore, Kr and Rb mathematical corrections were also accomplished by considering $^{86}Kr/^{83}Kr$ and $^{87}Rb/^{85}Rb$ ratios of 1.50566 and 0.38567, respectively.

All the isotopic values obtained for the NIST SRM 987 standard solution during each measuring session were used to evaluate the instrumental precision.

4.5. Statistical Analysis

The data comparison was performed using a Student *t*-test method and, in the case of more than two independent groups, by one-way analysis of variance (ANOVA). An alpha level of confidence of 0.05 was used for all statistical tests.

All statistical analyses were performed by using the "data analysis" macro of Microsoft® Excel®.

5. Results and Discussion

5.1. Influence of the Residual Rb on the Accuracy of the $^{87}Sr/^{86}Sr$ Ratio

Figure 1 reports the trend of the experimental data measured on the spiked solutions for the Sr ratio as a function of both the Sr and Rb concentrations. In particular, considering the NIST 987 Sr ratio value, 0.71026 ± 0.00002 (uncertainty (u) is expressed as twice the standard deviation (2sd) equivalent to the 28 ppm deviation calculated by Equation (1)).

$$\text{ppm} = \left(\frac{\Delta}{\text{NIST987}_{ratio}} \right) 10^6 \tag{1}$$

where Δ represents the difference between the measured and true Sr isotopic ratios. It is important to note that the trend of the data starts diverging as the Rb concentration increases. In fact, the mathematical correction of the isobaric interference on the $m/z = 87$ results almost still effective with a 25 µg kg^{-1} of Rb but only when Sr concentration is greater than 200 µg kg^{-1}.

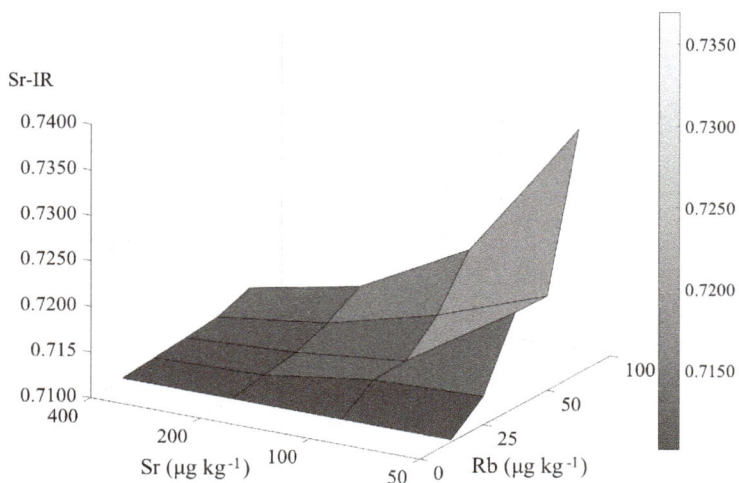

Figure 1. Values of $^{87}Sr/^{86}Sr$ determined on solutions varying both the Rb and Sr concentrations.

However, as long as the Sr concentration decreases from 400 to 50 µg kg^{-1}, the deviation of the Sr isotopic ratio from the expected value spans from 450 ppm to 4000 ppm. This means that the influence of the interference should become "negligible" or comprised inside the uncertainty of the NIST 987 standard when its concentration is approximately less than 5 µg kg^{-1} and the $^{87}Sr/^{86}Sr$ ratio is measured with an Sr concentration greater than 200 µg kg^{-1} [9].

Based on all these results, the measured isotope ratios were determined on SPE separated solutions at a final Sr concentration optimized for each sample at 200 µg kg^{-1}. The Rb content was measured for all the separated solutions and was always found to be less than 0.5 µg kg^{-1}.

5.2. Influence of the Milling Process

In order to test the potentialities of the milling process for the soil preparation, two different instruments were used. Table 1 reports the particle size distribution, expressed as percentage weight to weight, of the grinded soil samples obtained with a centrifugal mill, samples M1 and M2, and single jar planetary equipment, samples M3 and M4, respectively.

A detailed evaluation of the reported data immediately shows that both the milling techniques can produce a processed soil with a particle size distribution of less than 250 µm, and in particular, more than 80% of the sample is characterized by particles with diameters equal to or less than 125 µm. In addition, taking into account the values obtained from similar experiments—namely M1 versus M2 and M3 versus M4—both techniques show a good reproducibility.

Therefore, if on the one hand, the two milling techniques can operate on air-dried soil samples, more or less in the same way, then it is of utmost relevance to understand how similar, from a chemical point of view, are the grounded soils.

As a consequence, to investigate if the mill type may affect the Sr isotope ratio and then to give an answer to the latter question, the $^{87}Sr/^{86}Sr$ isotopic ratio, determined on the bioavailable Sr portion, was measured on each granulometric fraction, from 250 to <50 µm (collecting pan), relatively to four different soil samples processed with the planetary mill. Table 2 summarizes the isotopic ratio (IR) data.

Table 1. Comparison between the distribution size of soils milled by the centrifugal mill equipped with a 12 ribs rotor and 500 μm trapezoidal sieve ring, M1 and M2 samples, and the planetary mill equipped with a single agate jar, M3 and M4 samples.

Sample Size μm	Sample M1 %, w/w	Sample M2 %, w/w	Sample M3 %, w/w	Sample M4 %, w/w
315	0.10	—	—	—
250	3.00	4.00	5.90	—
180	6.90	13.00	12.20	14.40
125	16.10	17.10	13.60	21.50
90	26.80	28.00	19.40	14.20
63	32.10	29.20	25.80	26.00
50	9.90	7.30	12.40	11.30
Collecting pan	5.10	1.40	10.70	12.50

Considering each granulometric fraction, it is possible to highlight that the IR data are not influenced by a particle size effect, because the isotopic values are close to each other. In fact, the range of variability of the IR values of each sample lies within the uncertainties evaluated for the control sample (i.e., ±0.00002).

As far as the milling technique is concerned, Table 2 also reports the Sr isotopic values obtained on the same soil samples processed with the centrifugal mill (whole sample data).

Table 2. Values of the ^{87}Sr/^{86}Sr isotope ratio determined on the soils processed by the planetary mill at different particle sizes.

Sample Size μm	(^{87}Sr/^{86}Sr)_M1	(^{87}Sr/^{86}Sr)_M2	(^{87}Sr/^{86}Sr)_M3	(^{87}Sr/^{86}Sr)_M4
250	0.710088	0.709778	0.709179	0.711348
180	0.710081	0.709769	0.709149	0.711369
125	0.710103	0.709804	0.709183	0.711377
90	0.710105	0.709791	0.709183	0.711382
63	0.710090	0.709784	0.709194	0.711380
50	0.710092	0.709782	0.709192	0.711375
Collecting pan	0.710109	0.709763	0.709199	0.711366
Mean	0.71010	0.70978	0.70918	0.71137
SD	0.00001	0.00001	0.00002	0.00001
Whole sample [1]	0.71007	0.70978	0.70920	0.71136
|Δ$_{Mean-Whole\ sample}$|	0.00003	0.00000	0.00002	0.00001

[1] Value obtained on the soil samples processed by centrifugal mill equipment.

The calculated difference between the mean isotopic ratios and the whole sample data for the M1–M4 samples returns values that are of the same order of magnitude of the uncertainties of the data. This experimental evidence represents an important result when dealing with soil analysis, in particular ^{87}Sr/^{86}Sr ratio measurements, because this matrix may be considered a quite inhomogeneous system to be described by a punctual sampling procedure.

Concerning the determination of the ^{87}Sr/^{86}Sr ratio, the two mill types resulted in being equivalent. However, as the energy imparted to the sample is lower with the planetary grinding system and prevents spraying of materials, such as stones, bowls, and so on, this mill type was chosen as the pretreatment technique for the soils of the Chianti Classico area.

5.3. ^{87}Sr/^{86}Sr Ratio as Geographical Indicator in the Chianti Classico Production Area

With the aim to develop a preliminary geographical traceability model for wine, the isotopic ^{87}Sr/^{86}Sr values were determined on the soil, vine branch, and wine sampled from the Chianti Classico wine production zone. The results are reported in Table 3.

Table 3. Values of the ^{87}Sr/^{86}Sr ratio measured for soils, vine branches, and wines sampled from a producer in the Chianti Classico wine area.

Site	Hole	^{87}Sr/^{86}Sr Soil [1]	^{87}Sr/^{86}Sr Branch [2]	^{87}Sr/^{86}Sr Wine [3]
A	1up	0.71002	0.70963	
	1dw	0.70994		
	2up	0.70918	0.70914	0.70943
	2dw	0.70922		
Mean ± SD		0.70959 ± 0.00045	0.70939 ± 0.00035	
B	3up	0.70900	0.70905	
	3dw	0.70900		
	4up	0.70945	0.70924	
	4dw	0.70954		
	5up	0.70978	0.70977	
	5dw	0.71006		0.70941
	6up	0.70977	0.70941	
	6dw	0.71034		
	7up	0.70992	0.70929	
	7dw	0.70986		
Mean ± SD		0.70967 ± 0.00043	0.70935 ± 0.00027	

[1] Uncertainty value associated with the ^{87}Sr/^{86}Sr ratio data evaluated on the processed soil control sample $u = \pm 0.00002$ [20]. [2] Uncertainty value associated with the ^{87}Sr/^{86}Sr ratio data evaluated on the processed branch control sample $u = \pm 0.00002$ [20]. [3] Uncertainty value associated with the ^{87}Sr/^{86}Sr ratio data evaluated on the processed wine control sample $u = \pm 0.00002$ [20].

In particular, the soils were sampled from two distinct vineyards, A and B, and owing to the lithological pedological complexity of the investigated area, each soil core was split in an UP and DW parts.

The data do not point out any statistical difference between the respective UP and DW isotopic values for both A and B production sites. At the same time, the intra-site variability (within site A or B) is approximately one order of magnitude higher with respect to the inter-site variability (between sites A and B), confirming the possibility of considering the two production zones almost equal with respect to the ^{87}Sr/^{86}Sr indicator. In fact, by comparing the soil mean values for the ^{87}Sr/^{86}Sr ratio calculated for the A and B sites, these resulted in being statistically equivalent, $p(|t| \geq 0.32 = 0.75)$, making it possible to average all of the experimental IRs data relative to the whole investigated Chianti Classico wine production area.

Considering now the IRs determined on the vine branches, it is possible to highlight the following: (a) the isotopic ratio value measured on the vine branches is generally lower with respect to the respective soil datum, and (b) the comparison of the data measured on the samples from sites A and B confirms the similarities of the two production zones, because the A and B mean values are not statistically different, as confirmed by the results of the Student *t*-test, $p(|t| \geq 0.13 = 0.89)$.

Grapes from each vineyard were separately processed, and the Sr isotope ratios, determined on the fermented juice sample, were equal to 0.70943 and 0.70941 for wine_A and wine_B, respectively. Uncertainty values, expressed as twice the standard deviation, $u = 2SD$, were equal to ± 0.00002 [20].

Now, taking into account the overall mean value calculated for each investigated matrix—$IR_{soil} = 0.70965$, $IR_{branches} = 0.70936$, and $IR_{wine} = 0.70942$—it is possible to highlight that, although in some cases data are close to each other, the IR value decreases from soil to branches, while it was almost equal for vine branches and wine. In fact, the absolute differences of soil to branches (diff.$_{s-b}$ = 0.00029) is higher than that of branches to wine (diff.$_{b-w}$ = 0.00006), confirming that, in the latter case, there was an increase in the "representativeness" of the considered matrix and the discriminating capabilities of the isotopic indicator as well.

The same conclusions can be stressed by ANOVA analysis applied to the data of Table 3, namely the ^{87}Sr/^{86}Sr ratio of soils, vine branches, and wines.

The statistical analysis was performed on the values coming from site A and from site B considered as a whole on the basis of the evidence previously obtained from the Student *t*-test. Also, in this case, the ANOVA results reported in Table 4 confirm the absence of statistically significant differences between the means of the three groups.

Table 4. Results of the one-way ANOVA for statistically significant differences in the soils, vine branches, and wine data reported in Table 3.

Source of Variation	Sum of Squares	Degree of Freedom	F	F crit ($\alpha = 0.05$)
Between groups	4.17004×10^{-7}	2	1.530	3.492
Within groups	2.72386×10^{-6}	20		

This experimental evidence unequivocally confirms that it is possible to trace in an objective manner the geographical origin of food commodities following a "from farm to fork" approach. In particular, in absence of phenomena that may alter the soil–plant–product isotopic transfer mechanism, the vine branches return an isotopic ratio of strontium closer to that of the finished product, wine, because their sampling capacity and geographical representatively is far greater than that of any "operator" that makes holes. In addition, the vine branch sampling approach overcomes the great problem of the choice of the soil leaching test mimicking the evaluation of the bioavailable fraction and/or the seasonal dependence of the plant's uptake. Considering all these aspects, the main practical advantage of this approach is certainly represented by the increased discriminating capabilities of the isotopic indicator evaluated in the vine branch samples due to the reduced variability of the "within field data" attainable for non-homogeneous matrices such as soil.

6. Conclusions

The present investigation represents an important point in the development of a test method for determining the ^{87}Sr/^{86}Sr ratio in soil samples, in particular concerning the independence of the final data from the granulometric distribution of the sample. Regarding the effects produced by the two different grinding systems, despite the good correlation obtained between the granulometric distributions and the isotopic values with respect to the different tested samples, the use of a planetary system is preferable to a centrifugal one, because the lesser energy of impact on the soil sample prevents spraying of those materials, such as stones, bowls, and so forth, which do not directly participate in the plant element uptake processes. As far as the Rb interference is concerned, the reported data show that with only the mathematical correction, the calculated Sr isotope ratios are not sufficiently accurate, and therefore, the Rb/Sr separation process is mandatory.

The Sr isotopic values obtained for the Chianti Classico food chain support and confirm the idea that the values of the ^{87}Sr/^{86}Sr isotopic ratio in vine branches are very close to those in wine and therefore could be used for geographical traceability purposes in the oenological food chain.

Supplementary Materials: The following are available online at http://www.mdpi.com/2306-5710/4/3/55/s1, Table S1: Set-up parameters used for the PM100 planetary mill, Table S2: Microwave digestion program used for the mineralization of vine branches samples, Table S3: Microwave operating condition used for the washing cycle of XP-1500 Plus vessels, Table S4: ICP/qMS instrumental setting parameters, and Table S5: GPS coordinates for sampling sites A and B, respectively.

Author Contributions: Conceptualization, A.M. and C.D.; Methodology, S.S. and L.L.; Software, C.D.; Validation, S.S., L.L., and L.T.; Formal Analysis, C.D.; Investigation, S.S. and L.L.; Resources, L.T.; Data Curation, C.D. and L.L.; Original Draft Preparation, C.D. and A.M.; Review and Editing of Manuscript, C.D. and A.M.; Visualization, C.D. and S.S.; Supervision, A.M.; and Project Administration, A.M.

Funding: This research received no external funding.

Beverages **2018**, *4*, 55

Acknowledgments: The authors are grateful to the Centro Interdipartimentale Grandi Strumenti, University of Modena and Reggio Emilia, for their collaboration in working out the isotope ratio measurements with the Neptune instrument. A special thanks to Daniela Manzini and Maria Cecilia Rossi for their valuable contribution during the instrument setup and the helpful discussion on the measured data.

Conflicts of Interest: The authors declare no conflict of interest.

References

1. EU Agricultural Product Quality Policy. Available online: https://ec.europa.eu/agriculture/quality_en (accessed on 25 July 2018).
2. *Regulation (EU) 1151/2012 of the European Parliament and of the Council of 21 November 2012 on Quality Schemes for Agricultural Products and Foodstuffs*; L 343/1; Official Journal of the European Union: Brussels, Belgium, 14 December 2012.
3. Danezis, G.P.; Tsagkaris, A.S.; Camin, F.; Brusic, V.; Georgiou, C.A. Food authentication: Techniques, trends & emerging approaches. *Trends Anal. Chem.* **2016**, *85*, 123–132.
4. Baroni, M.V.; Podio, N.S.; Badini, R.G.; Inga, M.; Ostera, H.A.; Cagnoni, M.; Gallegos, E.; Gautier, E.; Peral-García, P.; Hoogewerff, J.; et al. How Much Do Soil and Water Contribute to the Composition of Meat? A Case Study: Meat from Three Areas of Argentina. *J. Agric. Food Chem.* **2011**, *59*, 11117–11128. [CrossRef] [PubMed]
5. Voerkelius, S.; Lorenz, G.D.; Rummel, S.; Quétel, C.R.; Heiss, G.; Baxter, M.; Brach-Papa, C.; Deters-Itzelsberger, P.; Hoelzl, S.; Hoogewerff, J.; et al. Strontium isotopic signatures of natural mineral waters, the reference to a simple geological map and its potential for authentication of food. *Food Chem.* **2010**, *118*, 933–940. [CrossRef]
6. Horn, P.; Schaaf, P.; Holbach, B.; Holzl, S.; Eschnauer, H. ^{87}Sr/^{86}Sr from rock and soil into vine and wine. *Z. Lebensm. Unters. Forsch.* **1993**, *196*, 407–409. [CrossRef]
7. White, W.M. *Geochemistry*; Wiley-Blackwell: Hoboken, NJ, USA, 2013; ISBN 978-0-470-65667-9.
8. Marchetti, A.; Durante, C.; Bertacchini, L. Heavy isotopes: A powerful tool to support geographical traceability of food. In *Food Authentication, Management, Analysis and Regulation*; Georgiou, C.A., Danezis, G.P., Eds.; Wiley Blackwell: Hoboken, NJ, USA, 2017; ISBN 9781118810255.
9. Fortunato, G.; Mumic, K.; Wunderli, S.; Pillonel, L.; Bosset, J.; Gremaud, G. Application of strontium isotope abundance ratios measured by MC-ICP-MS for food authentication. *J. Anal. At. Spectrom.* **2004**, *19*, 227–234. [CrossRef]
10. Aoyama, K.; Nakano, T.; Shin, K.C.; Izawa, A.; Morita, S. Variation of strontium stable isotope ratios and origins of strontium in Japanese vegetables and comparison with Chinese vegetables. *Food Chem.* **2017**, *237*, 1186–1195. [CrossRef] [PubMed]
11. Marchionni, S.; Buccianti, A.; Bollati, A.; Braschi, E.; Cifelli, F.; Molin, P.; Parotto, M.; Mattei, M.; Tommasini, S.; Conticelli, S. Conservation of ^{87}Sr/^{86}Sr isotopic ratios during the winemaking processes of 'Red' wines to validate their use as geographic tracer. *Food Chem.* **2016**, *190*, 777–785. [CrossRef] [PubMed]
12. Trincherini, P.R.; Baffi, C.; Barbero, P.; Pizzoglio, E.; Spalla, S. Precise determination of strontium isotope ratios by TIMS to authenticate tomato geographical origin. *Food Chem.* **2014**, *145*, 349–355. [CrossRef] [PubMed]
13. Walker, R.J.; Carlson, R.W.; Shirey, S.B.; Boyd, F.R. Os, Sr, Nd, and Pb isotope systematics of southern African peridotite xenoliths: Implications for the chemical evolution of the subcontinental mantle. *Geochim. Cosmochim. Acta* **1989**, *53*, 1583–1595. [CrossRef]
14. Price, T.D.; Burton, J.H.; Bentley, R.A. The characterization of biologically available strontium isotope ratios for the study of prehistoric migration. *Archaeometry* **2002**, *44*, 117–135. [CrossRef]
15. Willmes, M.; Bataille, C.P.; James, H.F.; Moffat, I.; McMorrow, L.; Kinsley, L.; Armstrong, R.A.; Eggins, S.; Grün, R. Mapping of bioavailable strontium isotope ratios in France for archaeological provenance studies. *Appl. Geochem.* **2018**, *90*, 75–86. [CrossRef]
16. Durante, C.; Baschieri, C.; Bertacchini, L.; Bertelli, D.; Cocchi, M.; Marchetti, A.; Manzini, D.; Papotti, G.; Sighinolfi, S. An analytical approach to Sr isotope ratio determination in Lambrusco wines for geographical traceability purposes. *Food Chem.* **2015**, *173*, 557–563. [CrossRef] [PubMed]

17. Sillen, A.; Hall, G.; Richardson, S.; Armstrong, R. [87]Sr/[86]Sr ratios in modern and fossil food-webs of the Sterkfontein Valley: Implications for early hominid habitat preference. *Geochim. Cosmochim. Acta* **1998**, *62*, 2463–2473. [CrossRef]

18. Durante, C.; Bertacchini, L.; Cocchi, M.; Manzini, D.; Marchetti, A.; Rossi, M.C.; Sighinolfi, S.; Tassi, L. Development of [87]Sr/[86]Sr maps as targeted strategy to support wine quality. *Food Chem.* **2018**, *255*, 139–146. [CrossRef] [PubMed]

19. Durante, C.; Baschieri, C.; Bertacchini, L.; Cocchi, M.; Sighinolfi, S.; Silvestri, M.; Marchetti, A. Geographical traceability based on [87]Sr/[86]Sr indicator: A first approach for PDO Lambrusco wines from Modena. *Food Chem.* **2013**, *141*, 2779–2787. [CrossRef] [PubMed]

20. Durante, C.; Bertacchini, L.; Bontempo, L.; Camin, F.; Manzini, D.; Lambertini, P.; Marchetti, A.; Paolini, M. From soil to grape and wine: Variation of light and heavy elements isotope ratios. *Food Chem.* **2016**, *210*, 648–659. [CrossRef] [PubMed]

21. Marchionni, S.; Braschi, E.; Tommasini, S.; Bollati, A.; Cifelli, F.; Mulinacci, N.; Mattei, M.; Conticelli, S. High-precision [87]Sr/[86]Sr analyses in wines and their use as a geological fingerprint for tracing geographic provenance. *J. Agric. Food Chem.* **2013**, *61*, 6822–6831. [CrossRef] [PubMed]

22. Braschi, E.; Marchionni, S.; Priori, S.; Casalini, M.; Tommasini, S.; Natarelli, L.; Buccianti, A.; Bucelli, P.; Costantini, E.A.C.; Conticelli, S. Tracing the [87]Sr/[86]Sr from rocks and soils to vine and wine: An experimental study on geologic and pedologic characterisation of vineyards using radiogenic isotope of heavy elements. *Sci. Total Environ.* **2018**, *628*, 1317–1327. [CrossRef]

23. Tessier, A.; Campbell, P.G.C.; Bisson, M. Sequential extraction procedure for the speciation of particulate trace metals. *Anal. Chem.* **1979**, *51*, 844–851. [CrossRef]

24. Albarède, F.; Telouk, P.; Blichert-Toft, J.; Boyet, M.; Agranier, A.; Nelson, B. Precise and accurate isotopic measurements using multiple-collector ICPMS. *Geochim. Cosmochim. Acta* **2004**, *68*, 2725–2744. [CrossRef]

25. Wieser, M.E.; Schwieters, J.B. The development of multiple collector mass spectrometry for isotope ratio measurements. *Int. J. Mass Spectrom.* **2005**, *242*, 97–115. [CrossRef]

26. Ingle, C.P.; Sharp, B.L.; Horstwood, M.S.A.; Parrish, R.R.; Lewis, D.J. Instrument response functions, mass bias and matrix effects in isotope ratio measurements and semi-quantitative analysis by single and multi-collector ICP-MS. *J. Anal. At. Spectrom.* **2003**, *18*, 219–229. [CrossRef]

27. Nowell, G.M.; Pearson, D.G.; Ottley, C.J.; Schwieters, J.; Dowall, D. Long term performance characteristics of a plasma ionization multi collector mass spectrometer (PIMMS): The Thermo Finnigan Neptune. In *Plasma Source Mass Spectrometry, Application and Emerging Technologies*; Holland, J.G., Tanner, S.D., Eds.; Royal Society of Chemistry: London, UK, 2003; pp. 307–320.

28. Ehrlich, S.; Gavrieli, I.; Dor, L.B.; Halicz, L. Direct high-precision measurements of the [87]Sr/[86]Sr isotope ratio in natural water, carbonates and related materials by multiple collector inductively coupled plasma mass spectrometry (MC-ICP-MS). *J. Anal. At. Spectrom.* **2001**, *16*, 1389–1392. [CrossRef]

29. Kim, R.J.; Yoon, J.K.; Kim, T.S.; Yang, J.E.; Owen, G.; Kim, K.R. Bioavailability of heavy metals in soils: definitions and practical implementation—A critical review. *Environ. Geochem. Health* **2015**, *37*, 1041–1061. [CrossRef] [PubMed]

30. Blum, J.D.; Taliaferro, E.H.; Weisse, M.T.; Holmes, R.T. Changes in Sr/Ca, Ba/Ca and [87]Sr/[86]Sr ratios between trophic levels in two forest ecosystems in the northeastern U.S.A. *Biogeochemistry* **2000**, *49*, 87–101. [CrossRef]

31. Aberg, G. The use of natural strontium isotopes as tracers in environmental studies. *Water Air Soil Pollut.* **1995**, *79*, 309–322. [CrossRef]

32. Aberg, G.; Fosse, G.; Stray, H. Man, nutrition and mobility: A comparison of teeth and bone from the Medieval era and the present from Pb and Sr isotopes. *Sci. Total Environ.* **1998**, *224*, 109–119. [CrossRef]

33. Bohlkea, J.K.; Horan, M. Strontium isotope geochemistry of ground waters and streams affected by agriculture, Locust Grove, MD. *Appl. Geochem.* **2000**, *15*, 599–609. [CrossRef]

34. Stein, M.; Starinsky, A.; Katz, A.; Goldstein, S.L.; Machlus, M.; Schramm, A. Strontium isotopic, chemical and sedimentological evidence for the evolution of lake Lisan and the Dead sea. *Geochim. Cosmochim. Acta* **1997**, *61*, 3975–3992. [CrossRef]

35. ISO. *Soil Quality—Extraction of Trace Elements from Soil Using Ammonium Nitrate Solution*; DIN ISO 19730 (2009-07); ISO: Geneva, Switzerland, 2009.

36. Cocchi, M.; Durante, C.; Marchetti, A.; Li Vigni, M.; Baschieri, C.; Bertacchini, L.; Sighinolfi, S.; Tassi, L.; Totaro, S. Optimization of microwave assisted digestion procedure by means of chemometric tools. In *Microwaves: Theoretical Aspects and Practical Applications in Chemistry*; Marchetti, A., Ed.; Transworld Research Network: Trivandrum, Kerala, India, 2011; pp. 203–223. ISBN 978-81-7895-508-7.

37. Baschieri, C. Food Traceability: A Multivariate Approach to Procedures Optimization and Models Development. Ph.D. Thesis, University of Modena and Reggio Emilia, Modena, Italy, 2012.

38. Berglund, M.; Wieser, M.E. Isotopic compositions of the elements 2009 (IUPAC Technical Report). *Pure Appl. Chem.* **2011**, *83*, 397–410. [CrossRef]

Article

Wine Traceability Using Chemical Analysis, Isotopic Parameters, and Sensory Profiles

Federica Bonello, Maria Carla Cravero *, Valentina Dell'Oro, Christos Tsolakis and Aldo Ciambotti

CREA Research Centre for Viticulture and Enology, Via Micca 35, 14100 Asti, Italy;
federica.bonello@crea.gov.it (F.B.); valentina.delloro@crea.gov.it (V.D.);
christos.tsolakis@crea.gov.it (C.T.); aldo.ciambotti@crea.gov.it (A.C.)
* Correspondence: mariacarla.cravero@crea.gov.it; Tel.: +39-01-4143-3814

Received: 21 May 2018; Accepted: 23 July 2018; Published: 27 July 2018

Abstract: NMR/IRMS techniques are now widely used to assess the geographical origin of wines. The sensory profile of a wine is also an interesting method of characterizing its origin. This study aimed at elaborating chemical, isotopic, and sensory parameters by means of statistical analysis. The data were determined in some Italian white wines—Verdicchio and Fiano—and red wines—Refosco dal Peduncolo Rosso and Nero d'Avola—produced from grapes grown in two different regions with different soil and climatic conditions during the years 2009–2010. The grapes were cultivated in Veneto (northwest Italy) and Marches (central Italy). The results show that the multivariate statistical analysis PCA (Principal Component Analysis) of all the data can be a useful tool to characterize the vintage and identify the origin of wines produced from different varieties. Moreover, it could discriminate wines of the same variety produced in regions with different soil and climatic conditions.

Keywords: NMR; IRMS; sensory analyses; traceability; geographical origin; isotopes

1. Introduction

European consumers are increasingly aware of wine quality and safety. Thus, the traceability of wines is surely a key topic. Legally speaking, the determination of the geographical origin of a wine product is an aspect strictly related to its certification, especially for authentic wines. The Common Market Organisation (CMO), which is the E.U. regulation on the processing and marketing of wines, introduced in 2011 two appellations aimed at harmonizing and integrating national laws and protocols: PDO (Protected Designation of Origin) and PGI (Protected Geographical Indications). In Italy, the production and commercialization of these wines represent an important sector of the country's economy. The new appellations, named DOP and IGP in Italian language, integrate the existing DOC, DOCG, and IGT. Moreover, autochthonous varieties have had a remarkable development in recent years. It follows that the enhancement of the methods used to verify the geographical origin of authentic wines is very important for their commercial success. Laboratories that provide controls have to develop more advanced methods of analysis able to find out frauds or to detect any contaminants present even in trace amounts [1]. Recently, some authors [2,3] proposed $\delta^{15}N$ as a further isotopic marker for the geographical characterization of grape products.

The strontium isotopic ratio 87Sr/86Sr can also be used for the traceability and authentication of wine [4]. In 2017, Moreira [5] showed that nanofiltration (NF), a membrane process with several applications in oenology, does not influence this isotopic ratio.

Isotopic techniques SNIF-NMR (Site Specific Natural Isotope Fractionation, Nuclear Magnetic Resonance) and IRMS (Isotopic Ratio Mass Spectroscopy) are now widely used to verify the origin of food and beverages. SNIF-NMR, a specialization of magnetic resonance, was developed in France by Gérard Martin [6] and his team in the late 1980s and was soon applied in the detection of wine frauds or nonconformities to law such as watering, sugaring, or the addition of exogenous sugars [7]. Subsequently, these methods were also used in the determination of the geographical origin of wines. To this aim, the variation of the following isotopic parameters are evaluated: isotopic data obtained by NMR techniques—parameter (D/H)I ppm—combined with those for IRMS—relations $\delta^{13}C‰$ vs. PDB $\delta^{18}O‰$ vs. SMOW and sensory analyses. In addition, isotopic ratios should be compared to reference databases.

The effectiveness of stable isotopes ratios (D/H) I, $^{13}C/^{12}C$, and $^{18}O/^{16}O$ in the assessment of the geographical origin of wines is affected by the natural variability of these parameters. Thus, their efficiency in the identification of wines improves if they are used jointly. (D/H)I and $^{18}O/^{16}O$ ratios depend on latitude but, in the meantime, $^{18}O/^{16}O$ is noticeably modified by the meteorological conditions during grape ripening. The most powerful ratios used to discriminate the areas of origin are (D/H)I and $^{18}O/^{16}O$ [7,8].

There is a wide range of literature on this subject. The most recent works [9–11] take into account the use of NMR-IRMS in the evaluation of climatic and geographical origin.

In addition, it is possible to integrate isotopic data with those obtained by sensory analyses. This can help to assess wine origin or distinguish wines from different areas. Only a few studies examine the possible relation between sensory data and isotope data. Rochfort et al. [12] showed that some mouth-feel parameters identified from sensory analysis can be usefully correlated to NMR-based metabolomics analysis of wine.

As part of a project on the geographical and varietal traceability of wines (QUALITEC-CRA), the relation among stable isotopes was evaluated in order to differentiate the areas of productions in northern and central Italy.

Thus, this work aimed at characterizing wines obtained by two white grape varieties, Fiano and Verdicchio, and two red grape varieties, Nero d'Avola and Refosco dal Peduncolo Rosso, cultivated in two regions, Veneto and Marches, located respectively in north and central Italy, during two vintages (2009–2010). Isotopic analyses (NMR and IRMS) and chemical and sensory analyses were determined to verify if together they could differentiate the wines of the same variety produced with grapes from the two production areas.

2. Materials and Method

The samples of Fiano and Verdicchio (white varieties) and Refosco dal Peduncolo Rosso and Nero d'Avola (red varieties) came from Veneto and Marches (vintages 2009 and 2010).

The vineyards had different soil characteristics. Veneto vineyards were situated in the volcanic area of Lessini Mountains (province of Vicenza) in Mason for Refosco dal Peduncolo Rosso and in Gambellara for Verdicchio; the Fiano vineyard was located in Piavon D'Oderzo (province of Treviso) in the high Venetian–Friulian plain whose soil is rich in carbonates. The one of Nero d'Avola was in a mulch–clay soil in the plain in Susegana (province of Treviso).

In Marches, all three vineyards were close to each other in the clay soils of the Loreto hills (province of Ancona), one for Nero d'Avola and Refosco dal Peduncolo Rosso, and the other two for Verdicchio and Fiano.

The climatic conditions in 2009 and 2010 were different, as reported in Table 1: the minimum and maximum average temperatures were higher, the average rainfall lower, and the evapotranspiration values higher in 2009 than in 2010 in both the regions. The corresponding average values of Italy were higher for temperatures and for the evapotranspiration values; for rainfall, they were lower with the exception of the average rainfall of Marches in 2009.

Table 1. Minimum and maximum average temperatures (°C), rainfall (mm), and evapotranspiration values (mm): average of 2009 and 2010 in Veneto, Marches, and Italy (data https://www.politicheagricole.it).

	Average Minimum Temperature (°C)		Average Maximum Temperature (°C)		Average Rainfall (mm)		Average Evapotranspiration (mm)	
	2009	2010	2009	2010	2009	2010	2009	2010
VENETO	7.3	6.7	16.6	15.4	904.3	1009.1	798.9	763.0
MARCHES	9	8.3	17.9	16.6	839.0	1036.7	876.3	828.8
ITALY	10.1	9.5	18.3	17.2	849.9	917.7	914.5	867.0

The grapes were vinified in duplicate (50 kg for each repetition) in the experimental cellar of the CREA Research Centre for Viticulture and Enology (Asti), according to the official protocols established by Reg. EC 555/2008.

2.1. White Wines Protocol

The white grapes were crushed and pressed with a hydraulic press, then the juice, with added SO_2 (30 mg/L) and pectolytic enzymes, was subjected to static defecation (12–15 °C) for about 18 h. After raking, the musts were inoculated with a commercial selected yeast starter (Lievito TOP15 produced by ENARTIS). The fermentation was carried out at 18–20 °C. The wines were added with SO_2 (50 mg/L), refrigerated for tartaric stabilization, clarified, filtered, and subsequently bottled.

2.2. Red Wines Protocol

The red grapes were crushed and destemmed, added with SO_2 (30 mg/L), and inoculated with selected yeasts (Uvaferm VRB). The maceration lasted 8–10 days with punching down twice daily. After racking, the wines were inoculated with bacteria *Oenococcus oeni* (Lalvin VP41) for the development of malolactic fermentation (MLF). The wines were then decanted, refrigerated for tartaric stabilization for 2 months, added with SO_2 (30 mg/L), and subsequently bottled in the February of the year after the harvest.

2.3. Analyses

All the samples of wines were submitted to chemical, physical, and isotopic analyses.

2.3.1. Chemical–Physical Analyses

The chemical–physical analyses, consisting of alcohol, dry extract, total acidity (g/L tartaric acid), volatile acidity (g/L acetic acid), and pH analysis, were performed according to standard methods (G.U. C.E. n. 272 3/10/1990).

Color intensity (E420 + E520) and hue (E420/E520) were measured in red wines. E420 was instead measured in white wines. The polyphenol indices were also determined according to the methodologies proposed by Di Stefano et al. [13].

2.3.2. Isotopic Analyses

Isotopic ratio (D/H) I (ppm) was measured by means of an NMR Bruker 300 MHz equipped with an automatic sample changer (Bruker BioSpin, Karlsruhe, Germany); other isotopic ratios were measured by means of a Mass Spectrometer Micromass VG Optima ISOTEC with an elemental analyzer Carlo Erba (NA 1500 NC, Carlo Erba, Milan, Italy) equipped with an automatic Fisons sampler AS-800 (Carlo Erba, Milan, Italy) to determine $\delta^{13}C$‰ vs. PDB and with a sampler ISOPREP 18 to analyze $\delta^{18}O$‰ vs. SMOW.

In order to verify the geographical origin of wines, an isotopic database on authentic wines was used, issued since 1987 under the support of the "Ministero delle Politiche Agricole e Forestali–Ispettorato Centrale per la Repressione delle Frodi".

2.3.3. Sensory Analyses

In the spring after vintage, the sensory analyses of the wines samples were performed by a panel of 12 expert judges (6 women and 6 men), all with extensive experience in wine sensory evaluation and belonging to the CREA Research Centre for Viticulture and Enology (Asti) technical and scientific staff.

The wines were identified with a 3-digit code, and the sensory sessions took place in a tasting room following ISO norms (8589-2007). Samples (50 mL) of the wines were evaluated in wine tasting glasses that follow ISO standard guidelines (3591-1977).

The sensory profiles were realized following a procedure derived from the ISO standards (11035-1994) and described in previous papers [14,15]. The descriptors to be included in a wine aroma wheel with unstructured scales were chosen from a predefined odor list [16] and integrated with visual and gustative descriptors. The choice of all sensory descriptors was performed during a preliminary tasting session. A frequency threshold for the attribute citations was established: the attributes of color, taste, and mouth-feel were chosen when their frequency of identification by the panel was greater than "(number of assessors × number of wines)/2". Regarding odor, its description is generally more complex: the 3rd-level descriptors were chosen when their frequency of identification was higher than "(number of assessors × number of wines)/4". All the selected attributes were confirmed and discussed by the panel with suitable standards (Table 2). A tasting sheet was created in order to measure the intensity of each chosen descriptor using an unstructured intensity scale presented on a wheel. Two replicates of samples of each wine were sensory analyzed.

Table 2. Reference materials for sensory analysis.

White Wines	
Attributes	**Reference Standards (per 300 mL Neutral-Flavored White Base Wine)**
acacia blossoms	essential oil (0.05 mL)
citrus	lemon fresh skin
apple	fresh apple infused
hexotic fruits	hexotic fruit juice (100 mL)
peach–apricot	peach juice (50 mL) and apricot juice (50 mL)
honey	honey (60 mL)
caramel	caramel (50 mL)
fresh herbaceous	3-cis-esanol (50 g)
aromatic herbs	mixture of dried aromatic herbs (10 g)
Red Wines	
Attributes	**Reference Standards (per 300 mL Neutral-Flavored White Base Wine)**
violet	β-ionone (0.3 μL)
berries	iced mixed berries (100 g)
cherry	cherry artificial aroma (0.05 mL)
plums	15–20 plums infused
marmalade	plum jam (60 g)
fresh herbaceous	3-cis-esanol (50 g)
Attributes	**Reference Standards (in Water)**
Acidity	citric acid (0.5 g/L)
Bitterness	caffeine (0.4 mg/L)
Astringency	alum (1 g/L)

2.3.4. Statistical Analyses

The results were subjected to ANOVA and the Tukey test (95%). The quantitative results of the sensory descriptors (average value of the evaluations of individual tasters) common to the white wines (Fiano and Verdicchio) or to the red wines (Refosco dal Peduncolo Rosso and Nero d'Avola) and the corresponding chemical and isotopic data were subjected to the statistical analysis with Principal Component Analysis (PCA) (XLSTAT 7.5).

3. Results

The isotopic parameters (Table 3), the chemical–physical results (Table 4), and the sensory analyses (Figures 1 and 2) constituted a dataset that allowed the study of the wine sample similarities through PCA (Figures 3–6).

Table 3. Isotopic parameters determined in white and red wine in the two vintages (2009–2010). The isotopic ratios are expressed in δ‰ relative to V-PDB (Vienna–Pee Dee Belemnite) for δ^{13}C, V-SMOW (Vienna–Standard Mean Ocean Water) for δ^{18}O. Different letters in the same column indicate significant differences at ANOVA and Tukey test (95%).

White Wines	(D/H)I	(D/H)II	R	δ^{13}C ‰ vs. PDB	δ^{18}O ‰ vs. SMOW
Fiano Marches 2009	100.12 cd	131.37 a	2.62 a	−28.78 a	4.53 b
Verdicchio Marches 2009	101.13 b	131.23 a	2.60 a	−26.96 c	4.52 b
Fiano Veneto 2009	100.83 bc	130.67 a	2.59 a	−28.07 b	4.70 b
Verdicchio Veneto 2009	102.30 a	130.33 a	2.55 b	−26.08 d	5.79 a
Fiano Marches 10	99.38 d	130.55 a	2.63 a	−28.69 a	1.68 c
Fiano Veneto 10	100.29 bcd	129.88 a	2.59 a	−27.96 b	1.40 c
Verdicchio Veneto 10	100.52 bc	127.05 b	2.53 b	−27.12 c	1.58 c
Pr > F	0.000	0.001	0.000	0.000	<0.0001
Significativity	yes	yes	yes	yes	yes
Red Wines	**(D/H)I**	**(D/H)II**	**R**	**δ^{13}C ‰ vs. PDB**	**δ^{18}O ‰ vs. SMOW**
Refosco Marches 2009	102.38 a	130.32 a	2.55 a	−25.71 a	3.16 a
Nero d'Avola Marches 2009	103.00 a	129.60 a	2.52 ab	−26.23 ab	3.48 a
Refosco Veneto 2009	102.42 a	128.07 ab	2.50 ab	−28.90 d	3.13 a
Nero d'Avola Veneto 2009	102.66 a	129.37 a	2.52 ab	−27.84 c	2.07 b
Refosco Marches 2010	101.16 a	129.95 a	2.57 a	−26.89 b	2.07 b
Nero d'Avola Marches 2010	101.37 a	129.10 a	2.55 a	−26.43 ab	2.08 b
Refosco Veneto 2010	101.62 a	125.99 b	2.45 b	−28.46 cd	0.14 d
Nero d'Avola Veneto 2010	101.64 a	128.29 a	2.52 ab	−28.25 cd	0.94 c
Pr > F	0.037	0.001	0.014	<0.0001	<0.0001
Significativity	yes	yes	yes	yes	yes

Table 4. Chemical–physical parameters determined in white and red wines in the two vintages (2009–2010). Total acidity is expressed as tartaric acid, volatile acidity is expressed as acetic acid, TP index is total phenol index. Different letters in the same column indicate significant differences at ANOVA and Tukey test (95%).

White Wines	Alcohol (% *V/V*)	Dry Matter (g/L)	pH	Total Acidity (g/L)	Volatile Acidity (g/L)	E420 (Color Intensity)
Fiano Marches 2009	14.36 a	23.05 b	3.11 c	8.40 ab	0.41 a	0.14 bc
Verdicchio Marches 2009	11.95 f	23.04 b	3.05 cd	8.25 ab	0.28 b	0.14 cd
Fiano Veneto 2009	13.62 c	21.9 bc	2.97 de	8.60 a	0.36 a	0.17 a
Verdicchio Veneto 2009	12.31 e	22.32 bc	3.55 a	6.20 d	0.22 bc	0.17 ab
Fiano Marches 10	14.05 b	21.00 c	3.07 c	6.85 c	0.23 bc	0.11 de
Fiano Veneto 10	14.28 a	22.55 bc	2.88 e	7.90 b	0.27 b	0.11 e
Verdicchio Veneto 10	13.01 d	25.50 a	3.41 b	6.45 cd	0.17 c	0.15 abc
Pr > F	<0.0001	0.001	<0.0001	<0.0001	<0.0001	0.000
Significativity	yes	yes	yes	yes	yes	yes
Red Wines	**Alcohol (% *V/V*)**	**Dry Matter (g/L)**	**pH**	**Total Acidity (g/L)**	**Volatile Acidity (g/L)**	**T P Index (mg/L)**	**E420/E520 (Color Hue)**	**E420 + E520 + E620 (Color Intensity)**
Refosco Marches 2009	12.68 a	23.50 bcd	3.75 ab	5.35 ab	0.40 b	1227 bc	0.795 b	0.585 b
Nero Marches 2009	11.54 abc	22.95 cd	3.15 d	6.05 a	0.36 b	1007 d	0.610 d	0.585 b
Refosco Veneto 2009	10.54 cd	21.30 e	3.48 bc	5.20 ab	0.61 a	924 d	0.745 bc	0.340 cd
Nero Veneto 2009	12.05 ab	22.40 de	3.29 cd	6.10 a	0.41 b	1091 cd	0.665 bcd	0.455 c
Refosco Marches 2010	11.74 abc	25.55 a	3.80 a	6.10 a	0.33 b	1370 ab	0.790 b	0.620 b
Nero Marches 2010	12.58 a	24.00 abc	3.66 ab	6.10 a	0.24 b	1484 a	0.640 cd	0.775 a
Refosco Veneto 2010	10.86 bcd	24.75 ab	3.89 a	4.95 b	0.42 b	1042 d	0.930 a	0.425 c
Nero Veneto 2010	9.83 d	21.00 e	3.66 ab	5.30 ab	0.29 b	1066 cd	0.745 bc	0.260 d
Pr > F	0.000	<0.0001	<0.0001	0.004	0.001	<0.0001	0.000	<0.0001
Significativity	yes	yes	yes	yes	yes	yes	yes	yes

The isotopic parameter (D/H) I (ppm) results (Table 3) showed significant differences among white wines of the same region (Fiano and Verdicchio 2009) or among the same wine of two vintages (Verdicchio Veneto). No differences were pointed out for this parameter among the red wines.

(D/H) II (ppm) showed similar value results in white and red wines in both regions.

A comparable situation was evidenced for the other isotopic parameters (R, δ^{13}C‰ vs. PDB and δ^{18}O‰ vs. SMOW): significant differences did not discriminate wine varieties or region of origin.

Table 4 shows the results of the chemical–physical analysis and the color parameters of white and red wines in 2009 and 2010.

The values of the different parameters—alcohol, dry matter, pH, total acidity, volatile acidity total polyphenols, and color measures—showed significant differences at ANOVA and Tukey test (95%). The significant differences did not discriminate wine varieties or region of origin.

Sometimes the significant differences were among wines of the same region or among the same wine of two vintages.

The sensory profiles of wines were different not only for each variety but also for the vintages (Figures 1 and 2).

For visual aspect, Verdicchio had golden highlights, while Fiano had yellow highlights. Some olfactory descriptors were common for Fiano and Verdicchio, such as acacia blossom and aromatic herbs. Fiano was characterized by the presence of notes of honey and citrus in both vintages. The 2009 wines were all characterized by aromatic herbs and apple notes.

The average sensory profiles of red wines (Figure 2) were more similar than white wines. For visual aspect, the color was ruby red with violet highlights for all the samples, but in 2010 there were high differences in the intensity of these descriptors in the two zones, as confirmed from chemical parameters. Many olfactory descriptors were common, such as violet, cherry, and berries. Refosco was characterized by the presence of notes of fresh herbaceous while Nero d'Avola by plums and marmalade. There were no differences in the descriptors for the two vintages.

The sensory data and the chemical and isotopic parameters were used for statistical elaboration.

The results of the PCA for white wines are reported in Figure 3, which shows the score plot for the first two principal components, which together explain 67.41% of the total variance in the two vintages (2009 and 2010).

The PCA shows a good separation of the samples based on vintage and on grape varieties. PC1 is able to separate the samples coming from the two white varieties, Fiano and Verdicchio. In particular, the samples of Verdicchio (at negative values) are well separated from the Fiano ones (at positive values). PC2, on the other hand, separates the samples according to their production year. Samples obtained from 2009 vintage are at positive values, while those from 2010 vintage are at negative values. Loadings plots analysis for PC1 and PC2 (Figure 4) evidence the most important variables for the samples distribution observed in the score plot.

Fiano wines were characterized by a greater acidity (lower pH), a yellowish color, and a better structure (body). On the other hand, Verdicchio wines had a greater softness, an intense acacia flowers fragrance, and a lower acidity. From an isotopic point of view, Verdicchio samples had a greater ^{13}C/^{12}C ratio. When considering the production year, 2009 samples had a greater SO$_2$ content and an increased ^{18}O/^{16}O ratio, while 2010 ones had a greater bitterness.

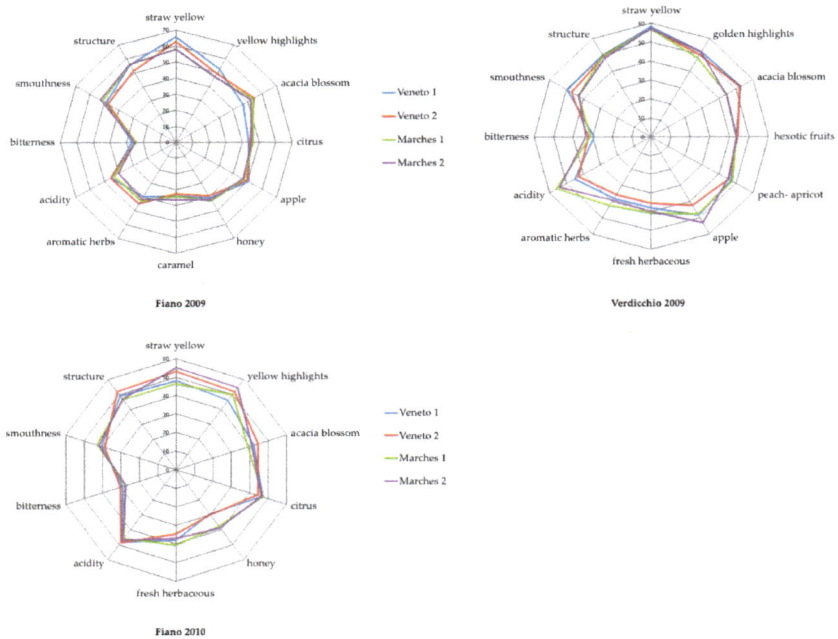

Figure 1. Sensory profiles of white wines.

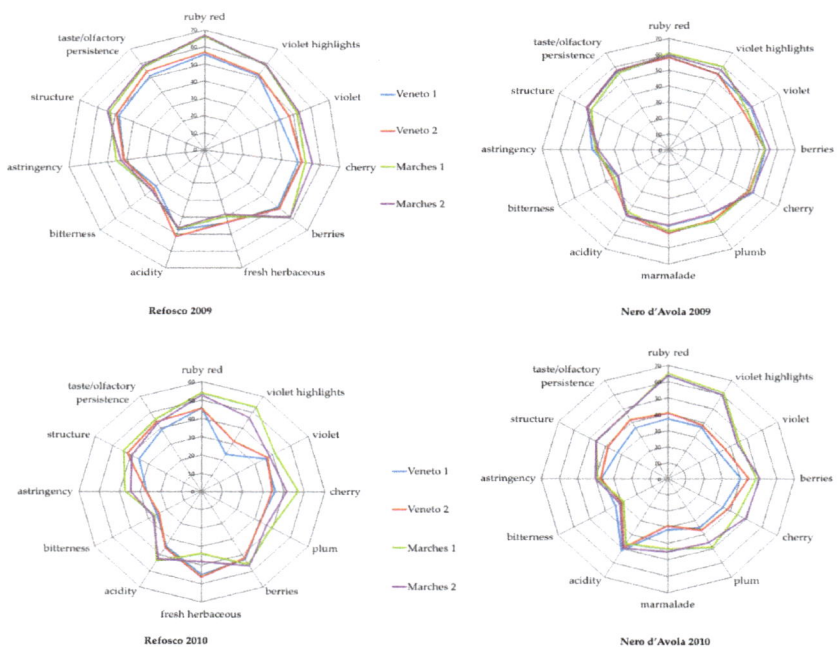

Figure 2. Sensory profiles of red wines.

Score Plot White wines F1 and F2: 67.41 %

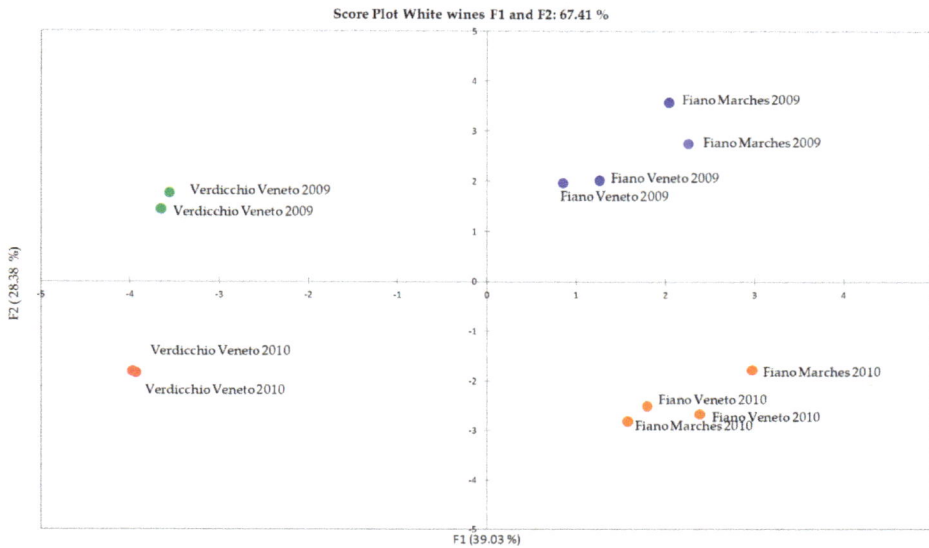

Figure 3. Score plot for white wines (Verdicchio and Fiano) in the two vintages (2009–2010).

Loadings plot Biplot F1 e F2: 67.41 %

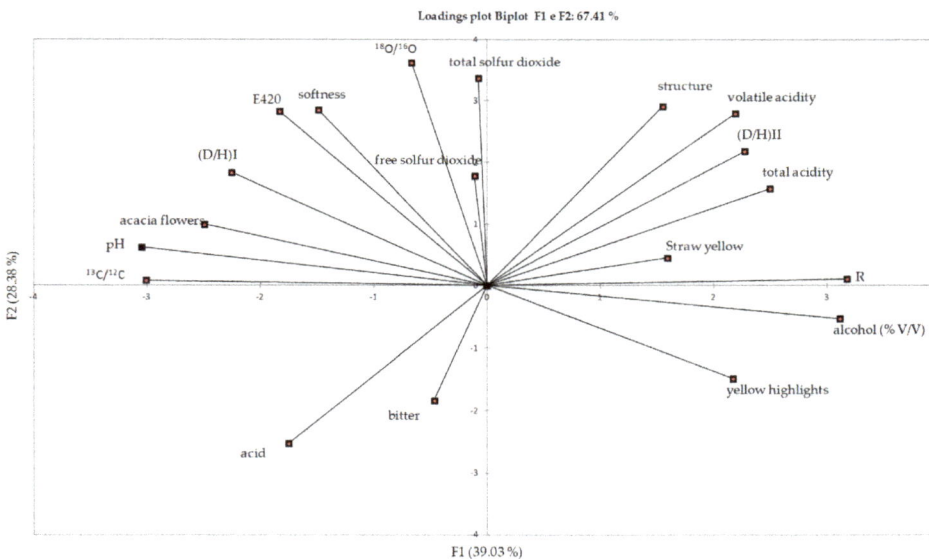

Figure 4. Loadings plot for white wines (Verdicchio and Fiano) in the two vintages (2009–2010).

Figure 5 shows the results of the PCA obtained considering all the data of the red wines. These samples are distributed along PC1 according to their production area while PC2 separates the two different vintages. The first two principal components together explain 60.50% of the total variance.

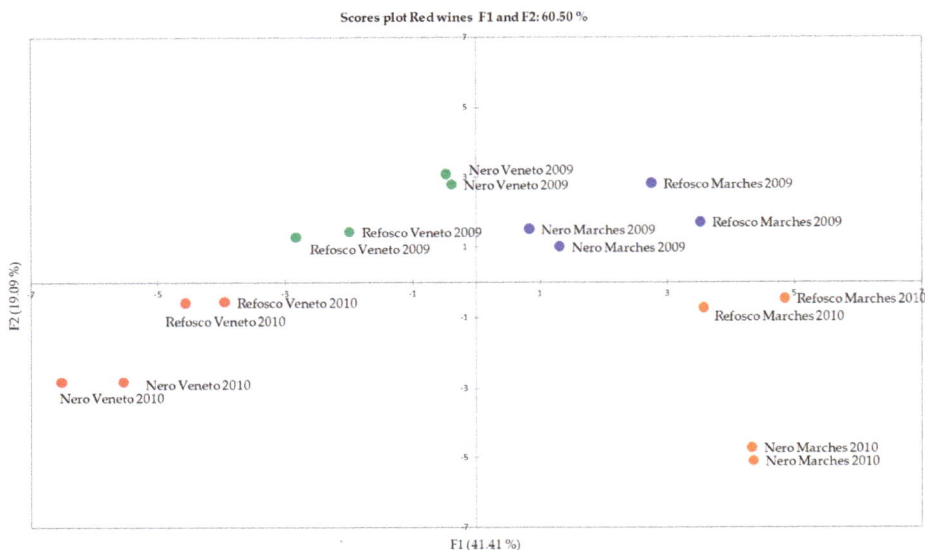

Figure 5. Score plot for red wines (Nero d'Avola and Refosco) in the two vintages (2009–2010).

Figure 6 shows the loadings plot that point out the parameters that most affect sample separation. Veneto samples showed a higher total acidity and color hue (E420/E520 ratio). Samples from 2009 vintage had a greater volatile acidity, bitterness, and lactic acid content. However, Marches samples were characterized by a major structure and persistence. These sensory parameters are correlated to the higher concentration of polyphenols, proanthocyanidins, and total flavonoids.

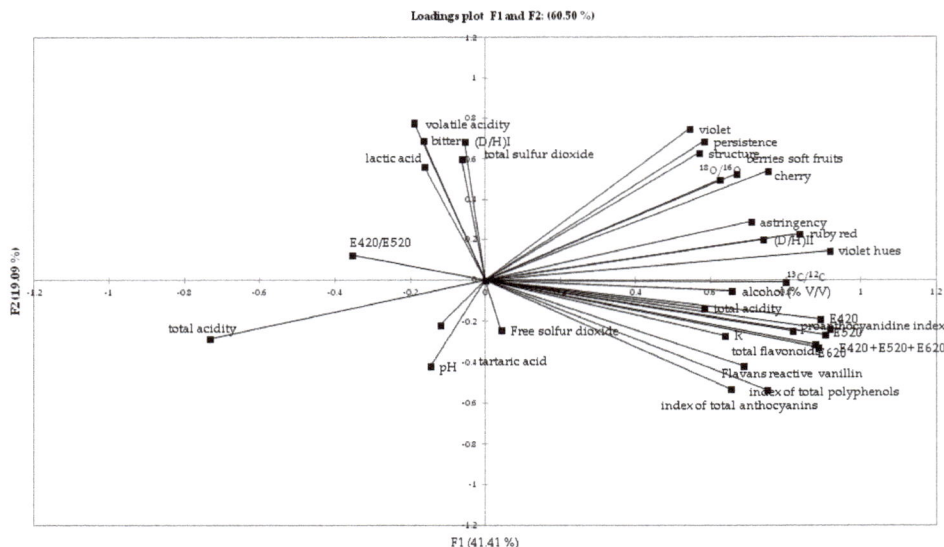

Figure 6. Loadings plot for white wines (Nero d'Avola and Refosco) in the two vintages (2009–2010).

4. Discussion

Isotopic parameters are successfully used to determine the geographical origin, a key aspect of wine traceability, with the help of the EU isotopic yearly data bank reports.

However, these ratios are influenced by the climatic conditions of the vintage. For example, in 2003, when the whole climate of Italy was characterized by the presence of high temperatures and low rainfall, the differentiation of the wine origin areas was not so clear [17]. According to [10], in vintages with particular conditions, like low rainfall or very high temperature, it is very difficult to differentiate the provenance of wines only with isotopic ratios.

In this experience, the isotopic results detected in the wines produced from grapes coming from Veneto and Marches were very much alike. Veneto climate is subcontinental (hot summers and cold winters) but it is mitigated by the sea and protected from northern winds by the Alps. As for Marches, in the areas by the sea, the climate is subcontinental too. Despite different latitudes, Veneto and Marches climates were quite similar in the years examined (Table 1).

The wine chemical data and the sensory analysis alone or together cannot discriminate the geographical origin. The vintage has generally a more significant effect than the origin of wines (Denomination of Origin) on the chemical and sensory data, as shown by [18] on 18 Spanish local varieties. A previous study on Nero d'Avola wines [14] also pointed out the effect of the vintage on these parameters.

The PCA multivariate analysis of the chemical, sensory, and isotopic parameters showed to be very helpful to discriminate the geographical origin of wines. The red samples are distributed along PC 1 according to their production area while PC 2 separates the two different vintages. In the white wines, there is a good separation of the samples based on vintage and on grape varieties.

5. Conclusions

This study shows that the application of the multivariate PCA to the isotopic data, coupled with chemical analyses and wine sensory profiles, can be more effective in the assessment of the geographical origin of the wines tested. The multivariate analysis gives a global vision on the distribution of wines according to the vintage and the origin. The number of wines in the study was limited and it could be interesting to extend the analysis on a bigger dataset.

Author Contributions: Conceptualization: V.D.; Methodology: M.C.C., A.C.; Formal analysis: F.B., C.T., A.C.; Data curation: F.B., M.C.C., C.T., A.C., V.D.; Writing—Original Draft Preparation: F.B., V.D., M.C.C.; Writing—review & editing: F.B., M.C.C.; Project administration: V.D.; Funding acquisition: V.D.

Funding: This research was funded by ENTECRA (now CREA, Council for Agricultural Research and Economics).

Acknowledgments: ENTECRA (now CREA) for the financial support and the sensory panel of the CREA Research Centre for Viticulture and Enology, in Asti (Italy) for the collaboration are gratefully acknowledged. The authors would like to thank Angelo Barrocu for collaborating on the analyses and Annamaria Di Franco for her help with the English text.

Conflicts of Interest: The authors declare no conflict of interest.

References

1. Camin, F.; Versini, G. Analisi innovative per marcare i prodotti lattiero-caseari di montagna. In *Alpeggi e Produzioni Lattiero Casearie: Atti del Convegno*; Regione Trentino-Alto Adige: Trento, Italy, 2001.
2. Durante, C.; Bertacchini, L.; Bontempo, L.; Camin, F.; Manzini, D.; Lambertini, P.; Marchetti, A.; Paolini, M. From soil to grape and wine: Variation of light and heavy elements isotope ratios. *Food Chem.* **2016**, *210*, 648–659. [CrossRef] [PubMed]
3. Santesteban, L.G.; Miranda, C.; Barbarin, I.; Royo, J.B. Application of the measurement of the natural abundance of stable isotopes in viticulture: A review. *Aust. J. Grape Wine Res.* **2015**, *21*, 157–167. [CrossRef]
4. Marchionni, S.; Braschi, E.; Tommasini, S.; Bollati, A.; Cifelli, F.; Mulinacci, N.; Mattei, M.; Conticelli, S. High-precision 87Sr/86Sr analyses in wines and their use as a geological fingerprint for tracing geographic provenance. *J. Agric. Food Chem.* **2013**, *61*, 6822–6831. [CrossRef] [PubMed]

5. Moreira, C.; de Pinho, M.; Curvelo-Garcia, A.S.; de Sousa, B.R.; Ricardo-da-Silva, J.M.; Catarino, S. Evaluating nanofiltration effect on wine 87Sr/86Sr isotopic ratio and the robustness of this geographical fingerprint. *S. Afr. J. Enol. Vitic.* **2017**, *38*, 82–93. [CrossRef]

6. Martin, G.J.; Guillou, C.; Martin, M.L.; Cabanis, M.T.; Tep, Y.; Aerny, J. Natural factors of isotope fractionation and the characterization of wines. *J. Agric. Food Chem.* **1988**, *36*, 316–322. [CrossRef]

7. Versini, G.; Monetti, A. Come e possibile controllare analiticamente l'origine geografica di un vino. *Enotecnico* **1996**, *32*, 77–89.

8. Gremaud, G.; Quaile, S.; Piantini, U.; Pfammatter, E.; Corvi, C. Characterization of Swiss vineyards using isotopic data in combination with trace elements and classical parameters. *Eur. Food Res. Technol.* **2004**, *219*, 97–104. [CrossRef]

9. Geana, E.I.; Popescu, R.; Costinel, D.; Dinca, O.R.; Ionete, R.E.; Stefanescu, I.; Artem, V.; Bala, C. Classification of red wines using suitable markers coupled with multivariate statistic analysis. *Food Chem.* **2016**, *192*, 1015–1024. [CrossRef] [PubMed]

10. Aghemo, C.; Albertino, A.; Gobetto, R.; Spanna, F. Correlation between isotopic and meteorological parameters in Italian wines: A local-scale approach. *J. Sci. Food Agric.* **2011**, *91*, 2088–2094. [CrossRef] [PubMed]

11. Dutra, S.V.; Adami, L.; Marcon, A.R.; Carnieli, G.J.; Roani, C.A.; Spinelli, F.R.; Leonardelli, S.; Vanderlinde, R. Characterization of wines according the geographical origin by analysis of isotopes and minerals and the influence of harvest on the isotope values. *Food Chem.* **2013**, *141*, 2148–2153. [CrossRef] [PubMed]

12. Rochfort, S.; Ezernieks, V.; Bastian, S.E.; Downey, M.O. Sensory attributes of wine influenced by variety and berry shading discriminated by NMR metabolomics. *Food Chem.* **2010**, *121*, 1296–1304. [CrossRef]

13. Di Stefano, R.; Cravero, M.C.; Gentilini, N. Methods for the study of wine polyphenols. *L'Enotecnico* **1989**, *25*, 83–89.

14. Cravero, M.C.; Bonello, E.; Tsolakis, C.; Piano, E.; Borsa, D. Comparison between Nero d'Avola wines produced with grapes grown in Sicily and Tuscany. *Ital. J. Food Sci.* **2012**, *24*, 385–387.

15. Guaita, M.; Petrozziello, M.; Motta, S.; Bonello, F.; Cravero, M.C.; Marulli, C.; Bosso, A. Effect of the closure type on the evolution of the physical-chemical and sensory characteristics of a Montepulciano d'Abruzzo Rosé Wine. *J. Food Sci.* **2013**, *78*, C160–C169. [CrossRef] [PubMed]

16. Guinard, J.X.; Noble, A.C. Proposition d'une terminologie pour une description analytique de l'arôme des vins. *Sci. Aliments* **1986**, *6*, 657–662.

17. Bonello, F.; Cravero, M.C.; Tsolakis, C.; Ciambotti, A. Applicazione dei metodi isotopici e dell'analisi sensoriale negli studi sull'origine dei vini. In Proceedings of the VIII International Terroir Congress, Soave, Italy, 14–18 June 2010; Volume 2, pp. 69–74.

18. García-Muñoz, S.; Muñoz-Organero, G.; Fernández-Fernández, E.; Cabello, F. Sensory characterisation and factors influencing quality of wines made from 18 minor varieties (*Vitis vinifera* L.). *Food Qual. Preference* **2014**, *32*, 241–252.

MDPI

Article

Wine Traceability with Rare Earth Elements

Maurizio Aceto [1,*], Federica Bonello [2], Davide Musso [1], Christos Tsolakis [1,2], Claudio Cassino [1] and Domenico Osella [1]

[1] Dipartimento di Scienze e Innovazione Tecnologica, Università del Piemonte Orientale, viale T. Michel, 11, 15121 Alessandria, Italy; davide.musso@uniupo.it (D.M.); chr.tsolakis@gmail.com (C.T.); claudio.cassino@uniupo.it (C.C.); domenico.osella@uniupo.it (D.O.)

[2] CREA Consiglio per la Ricerca in Agricoltura e l'Analisi dell'Economia Agraria, Centro di Ricerca per l'Enologia, via Pietro Micca, 35, 14100 Asti, Italy; federica.bonello@entecra.it

* Correspondence: maurizio.aceto@uniupo.it; Tel.: +39-0131-360265

Received: 5 February 2018; Accepted: 6 March 2018; Published: 12 March 2018

Abstract: The traceability of foodstuffs is now a relevant aspect of the food market. Scientific research has been devoted to addressing this issue by developing analytical protocols in order to find the link between soil and food items. In this view, chemical parameters that can act as soil markers are being sought. In this work, the role of rare earth elements (REEs) as geochemical markers in the traceability of red wine is discussed. The REE distribution in samples from each step of the wine making process of *Primitivo* wine (produced in Southern Italy) was determined using the highly sensitive inductively coupled plasma-mass spectrometry (ICP-MS) technique. Samples analyzed include grapes, must, and wine samples after every step in the vinification process. The resulting data were compared to the REE distribution in the soil, revealing that the soil fingerprint is maintained in the intermediate products up to and including grape must. Fractionation occurs thereafter as a consequence of further external interventions, which tends to modify the REE profile.

Keywords: ICP-MS; rare earth elements; wine; traceability

---

1. Introduction

In the changing face of today's global wine industry, producers of traditional, quality wines are experiencing increased competition from low-quality, low-cost products. To combat potential frauds, the European Union has issued several regulations to ensure the authenticity of products labeled as coming from a specified geographic origin and produced according to a particular method. In addition, recently developed scientific techniques allow the origin of agricultural foodstuffs to be traced by using physical and chemical measurements of samples taken at different points throughout the entire production process, from the soil to the final edible product [1,2]. To be effective, these techniques need to take into account numerous factors that might influence the product, such as the climate and soil type, or the use of fertilizers and pesticides, not to mention the wine-making practices. As an example, techniques based on DNA analysis, which is independent from environmental influences, can be considered for traceability, especially for Protected Designation of Origin (PDO) products [3,4].

Inorganic parameters, especially isotopes of heavy elements such as lead and strontium, often turn out to be the most suitable for tracing a product's origin [5]. The isotope ratios of these elements have shown to be powerful tracers, allowing products to be linked to a particular soil. Isotopes of light elements—hydrogen, oxygen, nitrogen, and sulfur—are reliable indicators of food authentication, but their ratios are too variable to serve as tracers of the soil where a product originated [6].

Metals are another important group of chemical tracers. Of these, the rare earth elements (REEs) are the most reliable, and several studies have shown that trace metals, particularly REEs, can act as geochemical markers [7–9]. However, relatively little information is available on the use of REEs in the traceability of foodstuffs. The high specificity of REEs derives from their compact grouping of 14 elements, from lanthanum to lutetium, with very similar chemical properties arising from their 4f electronic configuration. Another potential advantage of REEs is that they are much less heavily exploited in industry than other transition metals, and are thus not as ubiquitous in the natural environment. These characteristics make REEs useful as geochemical markers and attractive as agents for food traceability [10,11]. A further advantage of REEs is that they do not play a specific role in the metabolism of plants and are therefore taken up indiscriminately from the soil (although in diluted amounts) by the plant, with no fractionation of the original distribution. In fact, fractionation is an important consideration in the traceability study of a particular foodstuff: each passage in the production chain must be carefully inspected for whether or not it could induce fractionation in the original distribution of REEs.

Several studies have exploited the distribution of REEs and, more generally, of trace and ultra-trace elements to develop schemes of wines classification, as shown in recent reviews [12,13]. Most of these studies have been devoted to the *authentication* of wines, seeking to determine whether samples could be discriminated according to their geographical provenance [14], the variety of grape [15], or to oenological features such as ageing [16]. Very few studies have been devoted to the *traceability* of wine, in order to verify the link between a particular wine and the soil in which its grapes were grown. While authentication studies can be based on several types of chemical markers, e.g., trace elements, isotopic data [17,18], volatile compounds [19], and polyphenols [20], only the first two parameters have proven useful in traceability studies [21]. Hopfer et al. [22] in their study on Californian wines were able to classify the samples according to their vineyard origin and their processing winery, showing that the discrimination is possible according to both soil elemental content and viticultural practices. In one of the first studies on the traceability of wines, the soil–wine correlation in two wine-producing regions in Canada, Okanagan Valley, and the Niagara Peninsula was investigated [23]. They found that strontium was able to differentiate both soils and wines from the two regions. Another study analyzed four complete oenological production chains from Piedmont (Italy): *Gavi*, *Barbera*, *Brachetto d'Acqui*, and *Freisa* [24]. Samples of soil, grapes, must, and wine were analyzed with inductively coupled plasma-mass spectrometry (ICP-MS), giving particular care to REEs. The results showed that REE distribution was the same in the must as in the original soil, while fractionation occurred in the wine as a consequence of the winemaking process. Similar results were obtained on the production chain of a *Moscato d'Asti*: the distribution of REEs was the same in the soil, grapes, and must, but fractionation occurred in the wine after clarification with bentonites [25]. Significant correlation between soil and wine data sets were found in samples from three regions in Argentina, where both elemental composition and isotopic ratios ($^{87}Sr/^{86}Sr$) were used [26].

In this work, we further examined the role of REEs in the traceability of wine by analyzing the entire production chain of *Primitivo di Manduria* PDO wine (hereafter, *Primitivo*), a red wine produced in the area of Taranto and Brindisi, two main towns of the Apulia region, in Southern Italy. This wine has been awarded the Controlled Designation of Origin (CDO) mark according to an Italian Decree of 30 October 1974, and this mark was successively modified to a PDO. As a consequence, this wine is produced according to a stringent technical sheet, which specifies also the allowed production area. Samples, including soil, were taken at each step of production: grapes, must, and wine aliquots were taken following each step of vinification, including the process of fining in barriques (wooden barrels). Analysis of the grapes consisted in separate measurements taken on the pulp, skins, and seeds, so that the distribution of trace metals in different parts of the fruit could be compared.

Being REEs present at µg/L level or less in wine, a highly sensitive technique must be used in order to produce reliable data. We therefore employed inductively coupled plasma-mass spectrometry (ICP-MS), a multi-elemental technique with high sensitivity, accuracy, and precision.

2. Materials and Methods

2.1. Materials

High-purity water from a Milli-Q apparatus (Milford, MA, USA) was used in the study. 30% TraceSelect hydrogen peroxide, 69% nitric acid, 95% sulfuric acid, and 37% hydrochloric acid were purchased from Fluka (Milan, Italy). Polypropylene and polystyrene vials, used respectively for sample storage and analysis with an auto-sampler system, were kept in 1% nitric acid and then rinsed with high-purity water when needed. Porcelain capsules of a 30 mL volume were used for microwave dry ashing. Elements stock solutions (Inorganic Ventures, Lakewood, NJ, USA) were used for external calibration (La, Ce, Pr, Nd, Sm, Eu, Gd, Tb, Dy, Ho, Er, Tm, Yb, and Lu), stability testing (Li, Co., In, Ce, and U), and internal standardization (Rh, In, and Bi).

2.2. Sample Collection

A complete production chain was analyzed. Soil, grapes, musts, and wine samples were provided by a winery in Grottaglie (TA, Italy). At harvesting, 300 berries from a vineyard were picked from the upper, middle, and bottom parts of the bunches, from both shaded and exposed sides of the row, and then pooled. Sampling was carried out in August 2012. The collected berries were further divided into three groups of 100 berries, which were used as triplicates for the determinations. Grapes and must samples were kept frozen at $-20\ ^{\circ}$C in order to stop all fermentation reactions; they were thawed for 4 h before sample treatment.

2.3. Sample Treatment

Being of different chemical nature, the various samples were treated with different procedures as follows:

1. Soil samples were treated according to a standardized procedure: soil was dried at 105 $^{\circ}$C for 24 h, after which 1 g was sieved (ϕ 0.2 mm) and extracted with 20 mL of hydrogen peroxide for 20 min and then with 12 mL of aqua regia on a heating plate for 2 h under reflux. The resulting solution was diluted to volume in a 100 mL volumetric flask with high-purity water.

2. The grapes were processed as follows: skins were manually separated from the pulp and seeds and transferred to different Pyrex glass containers. All parts were separately subjected to dry ashing in porcelain crucibles in a Pyro 260 microwave ashing system (Milestone, Sorisole, Italy) with the following temperature cycle: 15 min to 150 $^{\circ}$C, 60 min to 1000 $^{\circ}$C, and 10 min at 1000 $^{\circ}$C. The resulting ash was dissolved in 1 mL of concentrated nitric acid and brought to 50 mL to obtain a nitric acid concentration of approximately 1%. All solutions were prepared with high-purity water.

3. Musts (100 g) were dried overnight at 105 $^{\circ}$C. The dried samples were transferred to porcelain crucibles and subjected to ashing with the following temperature cycle: 50 min to 750 $^{\circ}$C, 10 min at 750 $^{\circ}$C, 10 min to 900 $^{\circ}$C, and 10 min at 900 $^{\circ}$C. The resulting ash was dissolved in 1 mL of concentrated nitric acid and brought to 50 mL to obtain a nitric acid concentration of approximately 1%. All solutions were prepared with high-purity water.

4. Wine samples, obtained after every passage in the vinification, were treated with microwave ashing in the same condition used for must.

All samples (soil, grapes, must and wine) were analyzed in triplicate.

2.4. Vinification Processes

Once the grapes had been picked up and transported to the winery, they were treated with 20 mg/L of SO_2. The grapes were initially processed in a crusher-destemmer. The liquid formed after crushing and pressing is the must, which was transferred into "cold-soaked" tanks for two days, and the must was then inoculated with yeast, and fermentation began. It is noteworthy that the time needed for fermentation varies according to the type of grapes and the method used by the winemaker. In the case of *Primitivo* wine, which typically has a high concentration of sugars, fermentation was carried out for a month. After fermentation, the juice (now wine) was pressed away from the skins into a holding tank, where it sat for a few days to allow sediment and dead yeast cells to settle out (P12SV). At this stage, P12SV wine was transferred into oak barrels for aging, and malolactic fermentation took place (with *Oenococcus oeni* bacteria inoculation). At the end of malolactic fermentation, monitored by HPLC, a sample was analyzed (P12FIN). This experimental wine created from *Primitivo* grapes was neither filtered nor stabilized.

Red wines may be aged from several months to several years, depending on the type and quality of the wine desired. For this study, we also analyzed wine samples coming from different vintages produced by the same winery (P08, P09, P10, and P11).

2.5. ICP-MS Analysis

Elemental analysis was performed with a Thermo Scientific (Waltham, MA, USA) X-Series II inductively coupled plasma mass spectrometer. The instrument is equipped with a quartz torch with a Plasma Screen device, a quadrupole mass analyzer, a lens ion optics based on a hexapole design with a chicane ion deflector and a simultaneous detector with real-time multichannel analyzer electronics, operating in either analogue signal mode or pulse counting mode. A high-efficiency ESI APEX-Q nebulizer (Epond SA, Vevey, Switzerland) was used as a nebulization system. The instrument and accessories were PC-controlled by PlasmaLab software. The instrument and measurement parameters were as follows: forward power: 1400 W; nebulizer gas flow: 0.92 L/min; auxiliary gas flow: 1.00 L/min; plasma gas flow: 13.1 L/min; dual mode detection with peak jumping; dwell time: 10 ms; 25 sweeps; 3 replicates for a total acquisition time of 180 s; the isotopes used were ^{139}La, ^{140}Ce, ^{141}Pr, ^{146}Nd, ^{147}Sm, ^{153}Eu, ^{157}Gd, ^{159}Tb, ^{163}Dy, ^{165}Ho, ^{166}Er, ^{169}Tm, ^{172}Yb, and ^{175}Lu. Interferences were evaluated as follows: $CeO^+/Ce^+ < 1\%$ and $Ba^{2+}/Ba^+ < 1\%$. A stability test was performed before each analysis session by monitoring ^{7}Li, ^{59}Co, ^{115}In, ^{140}Ce, and ^{238}U masses and making sure precision was better than 2%: instrumental precision was better than 2% for the trace elements, while the overall uncertainty (involving both sample preparation and instrumental analysis), calculated on the basis of five genuine replicates, was better than 5%. Background signals were monitored at 5, 101, and 220 m/z to perform a sensitivity test on the above-reported analyte masses. External calibration was employed for quantification, using multi-elemental standards prepared at five concentration levels in the range of 10–10,000 ng/L, by diluting multi-elemental stock solutions (100 mg/L) in 1% nitric acid solution. Internal standardization was used to correct for instrumental drifts by monitoring signals from ^{103}Rh, ^{115}In, and ^{209}Bi isotopes, which were in-line added to all samples, standards, and blanks at a concentration level of 10 µg/L; responses from the three isotopes were interpolated to yield a better correction. Detection limits for the elements determined were in the range 1–10 ng/L, calculated as 3 times the background standard deviation.

2.6. Analysis of Certified Samples

To evaluate the performance and recovery of the proposed sample treatments, three certified standard materials were analyzed. BCR 668 (Mussel tissue) and BCR 670 (Duckweed) from IRMM were analyzed using the same ashing procedure used for the grapes and musts, while SRM 2586 (trace elements in soil containing lead from paint) from the National Institute of Standards and Technology (NIST) was analyzed according to the same treatment described for the soil samples. The results, detailed in Tables 1–3, showed good agreement between the certified and observed concentration values, as already reported in the literature [10].

Table 1. Analysis of BCR 668 certified biological material (mussel tissue).

Element	Certified Values (µg/Kg)	Uncertainty	Found (µg/Kg)	s.d. *
La	80	6	76.12	1.87
Ce	89	7	106.41	4.30
Pr	12.3	1.1	13.21	0.36
Nd	54	4	52.54	1.70
Sm	11.2	0.8	11.02	0.89
Eu	2.79	0.16	3.14	0.13
Gd	13	0.6	12.93	1.15
Tb	1.62	0.12	1.66	0.18
Dy	8.9	0.6	8.39	0.60
Ho	1.8 [1]	0.60 [1]	1.62	0.20
Er	4.5	0.5	4.27	0.31
Tm	0.48	0.08	0.60	0.02
Yb	2.8 [1]	0.5 [1]	3.05	0.47
Lu	0.389	0.024	0.59	0.04

[1] Indicative value. * s.d.: standard deviation

Table 2. Analysis of BCR 670 certified biological material (duckweed).

Element	Certified Values (µg/Kg)	Uncertainty	Found (µg/Kg)	s.d.
La	487	20	474.1	4.3
Ce	990	40	959	134
Pr	121	6	115.2	11.1
Nd	473	15	483.1	14.5
Sm	94	7	96.10	0.53
Eu	23.2	1.5	76.12	0.32
Gd	98	8	105.61	6.37
Tb	14	1.1	13.24	2.12
Dy	79	7	80.22	4.91
Ho	15.8	1.8	16.51	2.98
Er	44	2.8	46.11	1.62
Tm	5.7	0.7	5.23	1.13
Yb	40	4	43.32	6.14
Lu	6.3	0.5	7.11	0.93

Table 3. Analysis of SRM 2586 certified soil material (trace elements in soil containing lead from paint).

Element	Certified Values (mg/Kg)	Uncertainty	Found (mg/Kg)	s.d.
La	29.7	4.8	27.11	0.96
Ce	58	8	56.82	1.64
Pr	7.3 [1]		7.51	0.21
Nd	26.4	2.9	26.14	0.81
Sm	6.1 [1]		5.22	0.189
Eu	1.5 [1]		0.98	0.04
Gd	5.8 [1]		4.82	0.18
Tb	0.9 [1]		0.68	0.03
Dy	5.4 [1]		3.52	0.15
Ho	1.1 [1]		0.63	0.03
Er	3.30 [1]		1.86	0.03
Tm	0.5 [1]		0.23	0.02
Yb	2.64	0.51	1.31	0.04
Lu	[2]		0.15	0.02

[1] Indicative value; [2] Not determined.

2.7. Data Analysis

Multivariate data analysis was applied to compare the distribution of REEs in all samples, verifying the effect played by the oenological practices on the REE distribution. Data analysis and graphical representations were performed with XLSTAT v. 2012.2.02 (Addinsoft, Paris, France), a Microsoft Excel add-in software package.

3. Results and Discussion

As expected, the REE distributions in all samples followed the Oddo–Harkins rule, according to which even-numbered nuclides are more abundant than their odd-numbered counterparts; the distributions therefore show the typical saw tooth profile with decreasing abundances. The distribution shown in Figure 1a corresponds to the REEs detected in *Primitivo* vineyard soil. Promethium (Pm) has not been determined, given its extremely low concentration, but it is reported between neodymium and samarium in the graph, making it clear that the Oddo–Harkins rule is maintained.

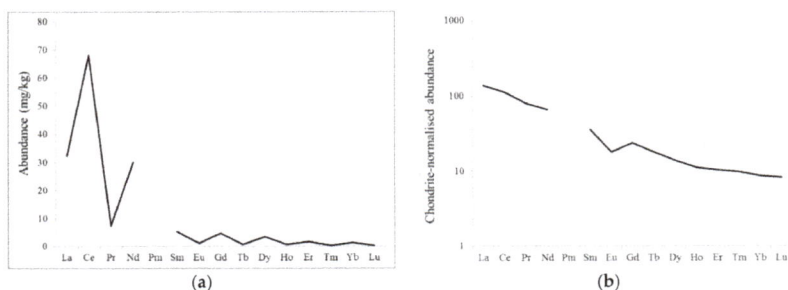

Figure 1. Distribution of rare earth elements (REEs) in *Primitivo* vineyard soil: (**a**) raw values; (**b**) chondrite-normalized values.

In geochemistry, it is common to compare data after normalization to typical patterns, such as those of *chondrite*, a meteoritic rock considered to be the best representative of the average concentrations of non-volatile elements in the solar system. Concentrations of single REEs in the samples are calculated according to Equation (1):

$$[REE]_{chondrite-normalized} = [REE]_{sample} / [REE]_{chondrite}. \tag{1}$$

Normalized data are displayed in logarithmic scale (Figure 1b). In this study, we used the values of a CI chondrite [27]. In addition, we employed an alternative normalization method by dividing REE concentrations, for every sample of the production chain, by the corresponding concentration of cerium (Ce)—the most abundant REE—according to Equation (2):

$$[REE]_{Ce-normalized} = [REE]_{sample} / [Ce]_{sample}. \tag{2}$$

We believe that this internal normalization to Ce is more suitable for comparing samples whose concentrations are of different orders of magnitude. The REE concentration in soil, in fact, is approximately three orders of magnitude higher than in must and even higher than in wine.

3.1. Distribution of REEs in Different Parts of the Grapes

The concentrations of REEs were determined separately in the pulp, skin, and seeds of the grapes, to compare their distribution. REE concentrations increased in the order seeds/pulp/skin (Figure 2). As expected, the distribution of REEs in must is very similar to pulp.

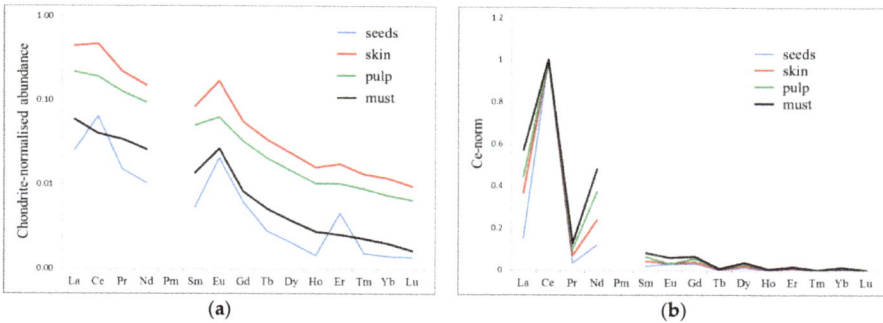

Figure 2. Distribution of REEs in the different parts of *Primitivo* grapes and in must: (**a**) chondrite-normalized values; (**b**) Ce-normalized values.

Unlike the other REEs, Europium (Eu) concentrations seemed to vary in the samples at every production step. Rather than to the geochemical behaviour of this element, the anomaly can be explained in terms spectral interference in ICP-MS analysis. In fact, $^{135}Ba^{16}O^+$ and $^{137}Ba^{16}O^+$ polyatomic ions can cause positive interference on ^{151}Eu and ^{153}Eu isotopes, respectively [28]; this interference cannot be solved with the instrumental resolution obtainable by the quadrupole mass analyzer used in this study. Barium, as a natural substitute of calcium, can be actively absorbed by plants, resulting in unpredictable positive interference with Eu.

3.2. Comparison of Soil and Must

Previous works, based on the geochemical behaviour of REEs in the *Vitis vinifera*/soil system [29–31] indicated that there is no fractionation of REEs in the passage from soil to grapes and from grapes to must [24,25]. It was therefore interesting at this stage to check whether the distribution of REEs in soil forms a sort of fingerprint maintained in must. In the *Primitivo* chain taken into consideration in this study, it is apparent from Figure 3a that the original fingerprint of soil is well maintained in the must. The resemblance of the REE distributions in the soil and must is even more apparent when shown as a Ce-normalized concentration (Figure 3b). The anomalous behaviour of europium can be explained as before: Ba^{2+} is relatively more abundant in must than in soil due to the fact that it is actively absorbed by the vine as a substitute of Ca^{2+}; therefore, its positive interference on Eu isotopes is also higher in must than in soil.

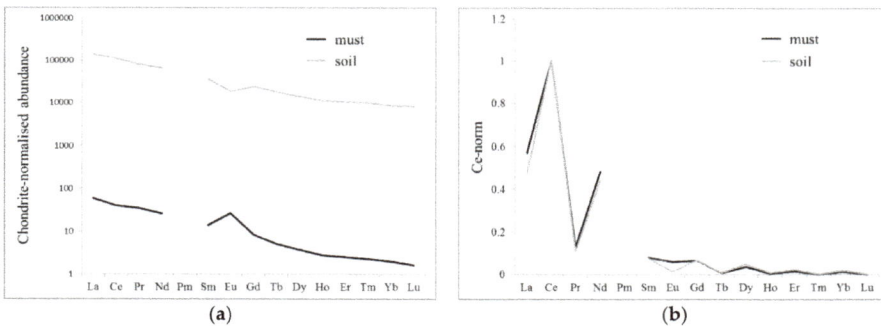

Figure 3. REE distributions in must and in soil: (**a**) chondrite-normalized values; (**b**) Ce-normalized values.

3.3. Effect of the Vinification Processes

The processes used for white winemaking cause fractionation of REE distribution, as obtained in the study on the traceability of a *Moscato* production chain [25]; this phenomenon is attributed to the release of REE ions from bentonites, which are clay materials widely used in oenology for wine clarification. Since bentonites are rarely used for red winemaking, the present study aims to determine whether REEs can act as suitable tracers all along the whole production chain of a red wine. In this case, two samples of wine were taken from the chain, one after alcoholic fermentation and another after malolactic fermentation. Both steps involve the addition of other substances: *Saccharomyces cerevisiae* yeasts were added to promote alcoholic fermentation, while *Oenococcus oeni* bacteria were added to promote malolactic fermentation. The possible contact of must/wine with surfaces, which can potentially release ions, should also be taken into consideration at this stage. The must was kept in stainless steel tanks until completion of alcoholic fermentation, and then the product was poured into barriques before malolactic fermentation. Figure 4 reports the REE distributions in the P12SV wine sample, withdrawn after alcoholic fermentation, and in the P12FIN sample, withdrawn after malolactic fermentation; the distributions in soil and must are included for comparison. Apart from the usual Eu anomaly explained by Ba interference, it seems evident that a certain degree of fractionation occurs after the alcoholic fermentation stage, making it hard to identify the original fingerprint given by the soil. The P12SV sample seems to be mainly depleted in Ce and Yb. Causes for this fractionation are attributable to the presence of inorganic additives in the biological products used for fermentation, including yeast nutrients, and in the release of metal ions from the surface of metallic tanks. The passage from P12SV to P12FIN seems less perceptible from a traceability point of view, possibly because storage in wooden barriques caused a negligible metal ions release.

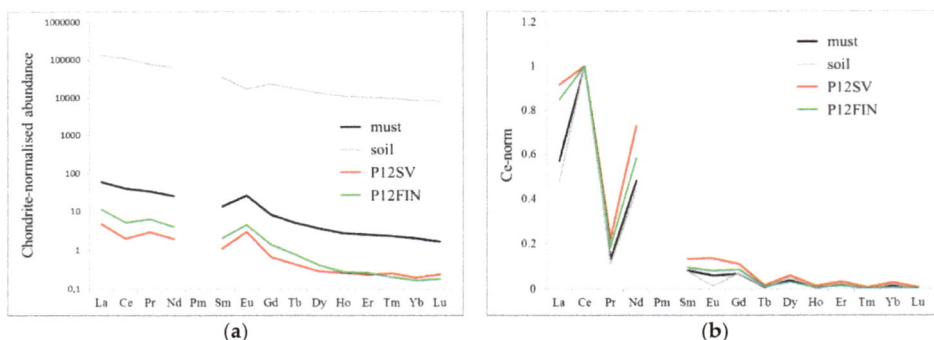

Figure 4. REE distributions in must, soil, and wines after fermentation processes: (**a**) chondrite-normalized values; (**b**) Ce-normalized values.

3.4. Effect of Vintage

In order to verify the effect of vintage, samples of *Primitivo* wines from different harvests were analyzed and compared. Samples labeled P08, P09, P10, and P11 come from the 2008, 2009, 2010, and 2011 harvests, respectively. Since *Primitivo* wine is aged one year in barriques before being bottled, it is obvious that Samples P08–P11 underwent a longer ageing process outside the cask than sample P12FIN, but we reputed that ageing in the bottle could at most cause only a phenomenon of precipitation of tartrates, which cannot induce fractionation in the distribution of REEs. When compared with each other and with the sample from the 2012 harvest (Figure 5), they show a high similarity. The P11 sample is rather anomalous, as far as La and Er concentrations are concerned.

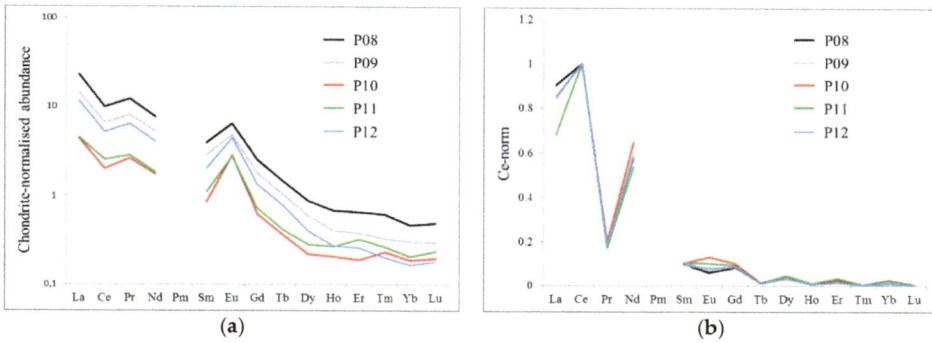

Figure 5. REE distributions in wines from different vintages: (**a**) chondrite-normalized values; (**b**) Ce-normalized values.

3.5. Multivariate Analysis

Multivariate chemometric analysis was performed on all data to verify the similarity between the various REE distributions. Agglomerative hierarchical clustering (AHC), using Euclidean distance as the similarity parameter and Ward's method as the agglomeration method, was selected for pattern recognition analysis. AHC was applied to Ce-normalized data after centering and scaling them; Ce and Eu were discarded as variables, the former being obviously equal to 1 in all samples and the latter being, in contrast, too variable due to barium interference. The result is displayed as a dendrogram in Figure 6: the similarity between *untreated* samples—the soil, must, and the various parts of grapes—is immediately apparent, while the *treated* samples, i.e., the wine samples, are grouped together in another cluster. The only exception is P11, which is grouped with the untreated samples; the reason for this behavior is due to its anomalous La and Er content, mentioned above.

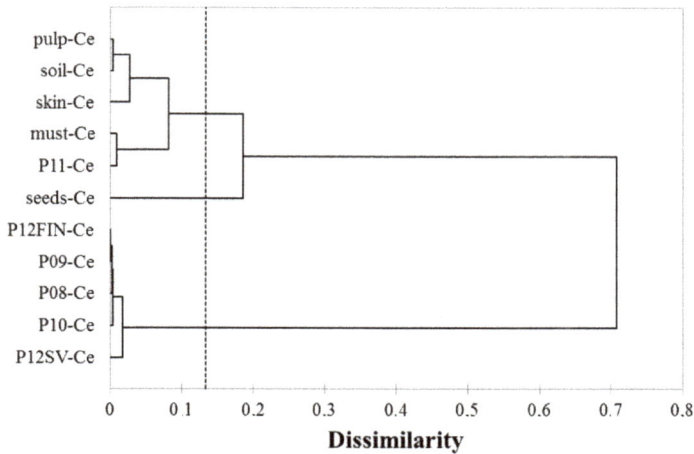

Figure 6. Dendrogram obtained with agglomerative hierarchical clustering (AHC) on Ce-normalized values.

4. Conclusions

REE distributions in samples taken at each step of a red wine making process showed that the original distribution in soil remains unaltered in every intermediate product up to and including the grape must. Variation of REE composition occurs thereafter as a result of additives used to promote fermentation or because of the interaction with the surfaces of storage tanks.

In order to fully evaluate the validity of a link between the original REE fingerprint of soil and a finished wine, additional study on the variations induced by technical interventions in the winemaking process will be useful. Moreover, a further study comparing different production chains of wines obtained from grapes grown on the same territory will be advisable, and in particular further investigations would be required by carrying out a more extended sampling plan, representative of the whole production area of the *Primitivo di Manduria* PDO wine.

Acknowledgments: The authors are grateful to the Regione Piemonte (Torino) (Marcandis and TRAMO projects) for financial support.

Author Contributions: M.A., F.B., and D.O. conceived and designed the experiments; F.B., D.M., C.C., and C.T. performed the experiments; M.A., F.B., and D.M. analyzed the data; M.A. and D.O. contributed reagents/materials/analysis tools; M.A. wrote the paper.

Conflicts of Interest: The authors declare no conflict of interest.

References

1. Furia, E.; Naccarato, A.; Sindona, G.; Stabile, G.; Tagarelli, A. Multielement fingerprinting as a tool in origin authentication of PGI food products: Tropea Red Onion. *J. Agric. Food Chem.* **2011**, *59*, 8450–8457. [CrossRef] [PubMed]

2. Benabdelkamel, H.; Di Donna, L.; Mazzotti, F.; Naccarato, A.; Sindona, G.; Tagarelli, A.; Taverna, D. Authenticity of PGI "Clementine of Calabria" by multielement fingerprint. *J. Agric. Food Chem.* **2012**, *60*, 3717–3726. [CrossRef] [PubMed]

3. Di Rienzo, V.; Miazzi, M.M.; Fanelli, V.; Savino, V.; Pollastro, S.; Colucci, F.; Miccolupo, A.; Blanco, A.; Pasqualone, A.; Montemurro, C. An enhanced analytical procedure to discover table grape DNA adulteration in industrial musts. *Food Control* **2016**, *60*, 124–130. [CrossRef]

4. Pasqualone, A.; Alba, V.; Mangini, G.; Blanco, A.; Montemurro, C. Durum wheat cultivar traceability in PDO Altamura bread by analysis of DNA microsatellites. *Eur. Food Res. Technol.* **2010**, *230*, 723–729. [CrossRef]

5. Drivelos, S.A.; Georgiou, C.A. Multi-element and multi-isotope-ratio analysis to determine the geographical origin of foods in the European Union. *TRAC Trends Anal. Chem.* **2012**, *40*, 38–51. [CrossRef]

6. Rossmann, A. Determination of stable isotope ratios in food analysis. *Food Rev. Int.* **2001**, *17*, 347–381. [CrossRef]

7. Brown, P.H.; Rathjen, A.H.; Graham, R.D.; Tribe, D.E. Rare earth elements in biological systems. In *Handbook on the Physics and Chemistry of Rare Earths*; Schneider, K.A., Eyring, L., Eds.; Elsevier: Amsterdam, The Netherlands, 1990; Volume 13, pp. 423–452, ISBN 978-0-444-88547-0.

8. Tyler, G. Rare earth elements in soil and plant systems: A review. *Plant Soil* **2004**, *267*, 191–206. [CrossRef]

9. Liang, T.; Ding, S.; Song, W.; Chong, Z.; Zhang, C.; Li, H. A review of fractionations of rare earth elements in plants. *J. Rare Earth* **2008**, *26*, 7–15. [CrossRef]

10. Oddone, M.; Aceto, M.; Baldizzone, M.; Musso, D.; Osella, D. Authentication and traceability study of hazelnuts from Piedmont, Italy. *J. Agric. Food Chem.* **2009**, *57*, 3404–3408. [CrossRef] [PubMed]

11. Aceto, M.; Musso, D.; Calà, E.; Arieri, F.; Oddone, M. Role of lanthanides in the traceability of the milk production chain. *J. Agric. Food Chem.* **2017**, *65*, 4200–4208. [CrossRef] [PubMed]

12. Versari, A.; Laurie, V.F.; Ricci, A.; Laghi, L.; Parpinello, G.P. Progress in authentication, typification and traceability of grapes and wines by chemometric approaches. *Food Res. Int.* **2014**, *60*, 2–18. [CrossRef]

13. Gonzálvez, A.; de la Guardia, M. Mineral Profile. In *Food Protected Designation of Origin: Methodologies and Applications*; de la Guardia, M., Gonzálvez, A., Eds.; Elsevier: Amsterdam, The Netherlands, 2013; Comprehensive Analytical Chemistry; Volume 60, pp. 51–76, ISBN 978-0-444-59562-1.

14. Marengo, E.; Aceto, M. Statistical investigation of the differences in the distribution of metals in Nebbiolo-based wines. *Food Chem.* **2003**, *81*, 621–630. [CrossRef]

15. Vaclavik, L.; Lacina, O.; Hajslova, J.; Zweigenbaum, J. The use of high performance liquid chromatography–quadrupole time-of-flight mass spectrometry coupled to advanced data mining

and chemometric tools for discrimination and classification of red wines according to their variety. *Anal. Chim. Acta* **2011**, *685*, 45–51. [CrossRef] [PubMed]

16. Marengo, E.; Aceto, M.; Maurino, V. Classification of Nebbiolo-based wines from Piedmont (Italy) by means of solid-phase microextraction-gas chromatography-mass spectrometry of volatile compounds. *J. Chromatogr. A* **2002**, *943*, 123–137. [CrossRef]

17. Almeida, C.M.R.; Vasconcelos, M.T.S.D. Does the winemaking process influence the wine [87]Sr/[86]Sr? A case study. *Food Chem.* **2004**, *85*, 7–12. [CrossRef]

18. Marchionni, S.; Braschi, E.; Tommasini, S.; Bollati, A.; Cifelli, F.; Mulinacci, N.; Conticelli, S. High-Precision Sr-87/Sr-86 analyses in wines and their use as a geological fingerprint for tracing geographic provenance. *J. Agric. Food Chem.* **2013**, *61*, 6822–6831. [CrossRef] [PubMed]

19. Capone, S.; Tufariello, M.; Francioso, L.; Montagna, G.; Casino, F.; Leone, A.; Siciliano, P. Aroma analysis by GC/MS and electronic nose dedicated to Negroamaro and Primitivo typical Italian Apulian wines. *Sens. Actuators B Chem.* **2013**, *179*, 259–269. [CrossRef]

20. Jaitz, L.; Siegl, K.; Eder, R.; Rak, G.; Abranko, L.; Koellensperger, G.; Hann, S. LC–MS/MS analysis of phenols for classification of red wine according to geographic origin, grape variety and vintage. *Food Chem.* **2010**, *122*, 366–372. [CrossRef]

21. Aceto, M. The use of ICP-MS in food traceability. In *Advances in Food Traceability Techniques and Technologies: Improving Quality throughout the Food Chain*; Espiñeira, M., Santaclara, F.J., Eds.; Woodhead Publishing: Sawston, UK, 2016; pp. 137–164, ISBN 978-00-8100-321-3.

22. Hopfer, H.; Nelson, J.; Collins, T.S.; Heymann, H.; Ebeler, S.E. The combined impact of vineyard origin and processing winery on the elemental profile of red wines. *Food Chem.* **2015**, *172*, 486–496. [CrossRef] [PubMed]

23. Taylor, V.F.; Longerich, H.P.; Greenough, J.D. Multielement analysis of Canadian wines by Inductively Coupled Plasma Mass Spectrometry (ICP-MS) and multivariate statistics. *J. Agric. Food Chem.* **2003**, *51*, 856–860. [CrossRef] [PubMed]

24. Aceto, M.; Baldizzone, M.; Oddone, M. Keeping the track of quality: Authentication and traceability studies on wine. In *Red Wine and Health*; O'Byrne, P., Ed.; Nova Science Publishers: New York, NY, USA, 2009; pp. 429–466, ISBN 978-1606927182.

25. Aceto, M.; Robotti, E.; Oddone, M.; Baldizzone, M.; Bonifacino, G.; Bezzo, G.; Di Stefano, R.; Gosetti, F.; Mazzucco, E.; Manfredi, M.; et al. A traceability study on the Moscato wine chain. *Food Chem.* **2013**, *138*, 1914–1922. [CrossRef] [PubMed]

26. Di Paola-Naranjo, R.D.; Baroni, M.V.; Podio, N.S.; Rubinstein, H.R.; Fabani, M.P.; Badini, R.G.; Inga, M.; Ostera, H.A.; Cagnoni, M.; Gallegos, H.; et al. Fingerprints for main varieties of Argentinean wines: Terroir differentiation by inorganic, organic, and stable isotopic analyses coupled to chemometrics. *J. Agric. Food Chem.* **2011**, *59*, 7854–7865. [CrossRef] [PubMed]

27. McDonough, W.F.; Sun, S.S. The composition of the Earth. *Chem. Geol.* **1995**, *120*, 223–253. [CrossRef]

28. May, T.W.; Wiedmeyer, R.H. A table of polyatomic interferences in ICP-MS. *Atom. Spectrosc.* **1998**, *19*, 150–155.

29. Censi, P.; Saiano, F.; Pisciotta, A.; Tuzzolino, N. Geochemical behaviour of rare earths in *Vitis vinifera* grafted onto different rootstocks and growing on several soils. *Sci. Total Environ.* **2014**, *473–474*, 597–608. [CrossRef] [PubMed]

30. Pisciotta, A.; Tutone, L.; Saiano, F. Distribution of YLOID in soil-grapevine system (*Vitis vinifera* L.) as tool for geographical characterization of agro-food products. A two years case study on different grafting combinations. *Food Chem.* **2017**, *221*, 1214–1220. [CrossRef] [PubMed]

31. Pepi, S.; Sansone, L.; Chicca, M.; Marrocchino, E.; Vaccaro, C. Distribution of rare earth elements in soil and grape berries of *Vitis vinifera* cv. "Glera". *Environ. Monit. Assess.* **2016**, *188*, 477. [CrossRef] [PubMed]

MDPI

St. Alban-Anlage 66

4052 Basel

Switzerland

Tel. +41 61 683 77 34

Fax +41 61 302 89 18

www.mdpi.com

Beverages Editorial Office

E-mail: beverages@mdpi.com

www.mdpi.com/journal/beverages

www.ingramcontent.com/pod-product-compliance
Lightning Source LLC
Chambersburg PA
CBHW051910210326

41597CB00033B/6099